Introduction to Coordination Chemistry

Edward Lisic, Ph.D.

Copyright © 2005 by Edward Lisic, Ph.D.

ISBN 0-7414-2702-8

Published by:

INFINITY
PUBLISHING.COM

1094 New DeHaven Street, Suite 100
West Conshohocken, PA 19428-2713
Info@buybooksontheweb.com
www.buybooksontheweb.com
Toll-free (877) BUY BOOK
Local Phone (610) 941-9999
Fax (610) 941-9959

Printed in the United States of America

Printed on Recycled Paper

Published June 2006

This book is dedicated to anyone and everyone who has ever synthesized a new ligand--- and then watched it react with a transition metal.

Thanks to: Dr. Jeff Wardeska, Dr. Brian Hanson, and Dr. Ed Deutsch — transition metal chemists, teachers and mentors.

Preface

After a departmental vote and years of discussion, our five-credit quantitative analysis class split into two classes, a three-credit quantitative analysis class and a two-credit sophomore inorganic class, CHEM2110. As the youngest inorganic chemist on the faculty, I was excited to take on the class.

Unfortunately, there was no available text for a 2-3 credit class at this level for coordination chemistry. The only other inorganic chemistry texts were prohibitively expensive for such a small enrollment course. I wanted a broad text on introductory transition metal chemistry, yet I wanted to impart to my students a love of the structures of these unique complexes and the flavor of their reactions with strange new ligands.

This text, in my opinion, is perfect for a sophomore level Transition Metal class, for freshman Honors classes or freshman inorganic chemistry classes, or as an inexpensive and concise supplement for advanced courses.

June 7th, 2006 Ed Lisic

Table of Contents

History of transition metal/inorganic chemistry

450BC Empedocles says that all matter is formed from **four elements - earth, air, fire** and **water**

400BC Democritus proposes that matter is actually composed of tiny indivisible particles, *atomos*

1661 Boyle's "The Sceptical Chymist" is published, introducing concepts of **elements, acids** and **alkalis**, and refuting many earlier claims by Paracelsus and ancient philosophers

1669 Hennig Brandt discovers **phosphorus** by distilling urine

1704 H. Diesbach creates **"Prussian Blue"** by accident, now known as $Fe_4[Fe(CN_6]_3 \cdot H_2O$

1766 Hydrogen discovered by **Henry Cavendish**

1774 Joseph **Priestley** discovers "dephlogisticated air", which **Lavoisier** renames **"oxygene"**

1774-89 Lavoisier proposes **Law of Conservation of Mass**

1798 Tassaert discovers $CoCl_3$ **x** $6NH_3$**,** one of the first known **coordination compounds**

1803 Dalton publishes table of **comparative atomic weights**

1805 Gay-Lussac proves that **water** is composed of two parts hydrogen to one part oxygen

1808 Dalton publishes his **atomic theory**

1813/14 Berzelius develops the **chemical symbols** and **formulas** used today

1817 Gmelin's Handbook of inorganic chemistry
1822 Gmelin prepares **cobalt ammonate oxalates**

1827 Zeise's salt discovered, $K^+[(C_2H_4)PtCl_4]$, the first olefin complex

1851 Genth, Claudet, Fremy prepare $CoCl_3 \cdot 6NH_3$, $CoCl_3 \cdot 5NH_3$, etc.

1852 Frankland proposes the concept of **Valence** (all atoms have a fixed valence)

1869 Blomstrand develops "**Chain theory**" of cobalt ammonates

1869 Dmitri Mendeleev publishes his first periodic table with 63 known elements

1884 Arrhenius proposes his **electrolytic** theory of **ions**

1890 Mond and **Bertholet** prepare $Ni(CO)_4$ and $Fe(CO)_5$

1892 Alfred Werner's dream about **coordination compounds**

1897 J. J. Thomson discovers that **electrons** (cathode rays) are negatively charged particles with very tiny mass
1898 Marie and Pierre Curie discover radium and polonium

1900 Max Planck proposes **Quantum Theory of electromagnetic radiation**

1900 Francois **Grignard** discovers magnesium **Grignard reagents**

1902 G.N. Lewis utilizes the first primitive **Lewis electron dot structures** in class (<u>cubic structures</u>)

1905 Alfred Werner's paper on **coordination chemistry** that leads to **Nobel Prize**

1905 Einstein explains the **photoelectric effect of metals,** and the **wave /particle duality of light**

1907 Werner synthesizes both **isomers of $CoCl_3 \cdot 4NH_3$**

1908 Fritz Haber reveals the **Haber Process** for producing ammonia from nitrogen and hydrogen

1909 The **pH scale of acidity** is devised by **Soren Sorensen**

1911 Optical isomers of *cis*-$[CoCl(NH_3)(en)_2]X_2$ resolved by **Werner**

1912 Max von Laue studies **X-ray diffraction** patterns and shows that crystals are repeating units of atoms

1913 Bohr model of the atom published (proposes the quantization of the energies of electrons)

1913 Werner wins first Nobel Prize given to an **inorganic chemist**

1913/14 The **Periodic Table** in its **present form** was drawn up by **Henry Moseley**

1919 Rutherford discovers that elements can be **transmutated**

1923 G.N. Lewis proposes **Lewis Acid/Base Theory and uses Lewis Electron Dot Structures**

1925 Fischer-Tropsch process is developed

1926 Shrödinger proposes the quantum-mechanical atom (electrons in orbitals about nucleus; electron spectroscopy explained as transitions among orbitals)

1927 Lewis ideas applied to electronic interpretation of coordination compounds by **Sidgwick**

1929 Crystal Field Theory proposed by **Bethe**

1930 Ziegler and Gilman simplify organolithium preparation, using ether cleavage and alkyl halide metallation respectively

1932 First application of Crystal Field Theory by Van Vleck

1935 Van Vleck combines Crystal Field Theory with Molecular Orbital Theory to produce what is now called the **Ligand Field Theory**

1937 Discovery of **Technetium** by **Emilio Segre**

1939 Linus Pauling publishes "The Nature of the Chemical Bond" which is in essence a summing up of seven papers of his published between 1931 and 1933

1939-40 Sidgwick proposes VSEPR Theory

1940 McMillan and Abelson produce the first transuranic element, neptunium

1941 Plutonium synthesized by **McMillan** and **Glenn Seaborg**

1951 Kealy, Pauson, Miller obtain **ferrocene $(C_5H_5)_2Fe$** first sandwich complex is discovered

1951 Orgel, Pauling, and Zeiss describe the **bonding in metal carbonyls**

1952 Henry Taube describes classical **Inner-Sphere and Outer-Sphere electron transfer processes**

1953 Karl **Ziegler** develops the first catalyst to convert monomers into polymers

1954 Linus Pauling wins the **Nobel Prize** for Chemistry for his work on chemical bonding

1955   Ziegler and Natta develop **olefin polymerization** at low pressure using mixed metal catalysts (transition metal halide / AlR_3)

1956 Review by **Moffit** and **Ballhausen** leads to acceptance of Crystal Field Theory by inorganic chemists

1958 The x-ray structure of $[CpMo(CO)_3]_2$ reveals a **metal-metal bond**

1959 Shaw and Chatt describe an **Oxidative-Addition reaction**

1961 Dorothy Hodgkin works out the structure of **vitamin B12** using a computer, which shows a cobalt-carbon bond

1961 Heck and Breslow determine the cobalt-carbonyl catalyzed **hydroformylation** pathway

1961 L. Vaska discovers that $(PPh_3)_2Ir(CO)Cl$ reversibly binds O_2 (**Vaska's Complex**)

1962 Xenon hexafluoroplatinate, the first compound of an inert gas, synthesized by Neil **Bartlett**

1963 Karl **Ziegler** and Giulio **Natta** awarded Nobel Prize for chemistry of high polymers

1964 Fischer isolates the first **metal-carbene** complex

1965 Wilkinson and Coffey, discover **(PPh$_3$)$_3$RhCl (Wilkinson's catalyst)** as a **homogeneous hydrogenation catalyst** for the hydrogenation of alkenes

1965 Cotton describes the **first quadruple metal-metal bond (delta bond) in [Re$_2$Cl$_8$]$^{2-}$**

1965 Rosenberg discovers that platinum compounds inhibit cell division, which leads to discovery of **cis-platin** as an anti-cancer drug

1968 First **organouranium** compound, **(C$_8$H$_8$)$_2$U**

1971 Yves Chauvin was able to explain in detail how metatheses reactions function and what types of metal compound act as catalysts in the reactions

1973 Fischer and Wilkinson win the **Nobel Prize** in Chemistry for synthesis and discovery of organometallic compounds (sandwich type)

1974 Cotton and co-workers discover the first **agostic complex** with a metal---H-C interaction

1979 The **Monsanto acetic acid process** from CO and methanol developed in the 1970's is described.

1980 E.A. Deutsch discovers that cationic complexes of [^{99m}Tc(diars)$_2$X$_2$]+ and [^{99m}Tc(dmpe)$_2$Cl$_2$]+ exhibit **myocardial uptake** ---thrusts **Tc** into **nuclear medicine** prominence

1982 Monsanto asymmetric hydrogenation process (Knowles etc.)

1983 Nobel Prize was awarded to **Henry Taube** for his work on the mechanisms of **electron transfer reactions** in metal complexes (Inner sphere and outer sphere)

1990 Richard Schrock produces the first efficient metal-compound catalyst for metathesis with metal carbenes

1992 Robert Grubbs develops a better metathesis catalyst which is stable in air. Commercial Ru catalysts have found many applications in organic chemistry (**Grubbs catalyst)**

2001 Nobel Prize was awarded to **Knowles, Noyori, Sharpless,** for work on **chirally catalyzed hydrogenation reactions** (Knowles & Noyori) and on chirally catalyzed oxidation reactions (Sharpless)

2005 Nobel Prize was awarded to **Yves Chauvin, Richard Schrock,** and **Robert Grubbs** for **metathesis catalysts**

1990-2000's
New generations of catalysts, supramolecular assemblies, molecular wires, and bioinorganic systems

Chapter 1. Coordination Chemistry: The Early Years

Until mankind retreated from the non-rational philosophies of *mysticism* that characterized the middle ages, the pseudo-science of *alchemy* was doomed to failure. Eventually, the adoption of the *scientific method* allowed humanity to break off our shackles of non-objective irrationality and advance into the modern age of science and reason.

Currently, with the accomplishments and discoveries of over two centuries of hard science at our command, we are expanding the scope and dimension of scientific discoveries at an unbelievable rate. Driven by the thirst for knowledge and the prize of economic incentives, our global culture is set to accomplish incredible things in the years to come.

The history of coordination chemistry is a case study of this process. Alchemy set its sight on the goal of transmutation of the elements into nobler substances, such as gold, but the use of non-rational procedures, magic, and mysticism clouded the real advances in technology and discovery that were made by our predecessors. Now, with the recent advances in organometallics, nano-technology, and bioinorganic chemistry, the future has never looked brighter for the science of inorganic chemistry.

The History of Coordination Compounds

Coordination compounds are typically characterized by four or six ligands (from the Latin word *ligare*, which means "to bind") in a coordination sphere bonded to a metal atom or ion in a tetrahedral, square planar, or octahedral geometry. This initial chapter starts an introductory investigation of coordination chemistry by putting its history into perspective and introducing some typical ligands and metal complexes.

The discovery of coordination compounds and the subsequent bonding explanations should be viewed against the larger history of progress in chemistry in understanding atomic structure of atoms, the periodic table of elements, and chemical bonding.

Coordination compounds were used for years before their true compositions and structures were determined. Heinrich **Diesbach** created **"Prussian Blue"** by accident in1704, which is now known as $Fe_4[Fe(CN_6)]_3 \cdot H_2O$, by mixing contaminated potash with iron sulfate. Instead of red, which he was expecting, he got purple, then a deep blue when it was concentrated. This is the first, or one of the first, known coordination compounds. It could also be considered to be the first organometallic species synthesized.

Joseph Proust is best known in connection with a long controversy with **C. L. Berthollet** who was led by his doctrine of mass-action to deny that substances always combine in constant and definite proportions. Proust maintained that compounds always contain definite quantities of their elements. In 1799 he proved that carbonate of copper, whether natural or artificial, always has the same composition, and later he showed that the two oxides of tin and the two sulfides of iron always contain the same relative weights of their components and that no intermediate indeterminate compounds exist.

Antoine Lavoisier published *Reflexions sur le Phlogistique* (1783), where he showed the phlogiston theory to be inconsistent. In *Methods of Chemical Nomenclature* (1787), he invented the system of chemical nomenclature still largely in use today, including names such as sulfuric acid, sulfates, and sulfites. His *Traité Élémentaire de Chimie (Elementary Treatise of Chemistry,* 1789) was the first modern chemical textbook, and presented a unified view of new theories of chemistry, contained a clear statement of the Law of Conservation of Mass, and denied the existence of phlogiston. In addition, it contained a list of elements, or substances that could not be broken down further, which included oxygen, nitrogen, hydrogen, phosphorus, mercury, zinc, and sulfur.

Contributions by Joseph Proust (***Law of Definite Proportions*** which is also called the Law of Constant

Composition) and Antoine Lavoisier (father of modern chemistry - *Law of Conservation of Mass*), along with many others, led John Dalton to formulate the first atomic theory.

John Dalton (1766-1844) developed the first useful *atomic theory of matter* around 1803. In his studies on meteorology, Dalton concluded that evaporated water exists in the air as an independent gas. Dalton reasoned that if water and air were composed of discrete individual particles, then evaporation might be viewed as a mixing of water particles with air particles. Dalton developed the hypothesis that the *sizes* of the particles making up different gases must be different while trying to explain the results of his experiments. In **1808** he published his *New System of Chemical Philosophy.*

Throughout the 1st half of the nineteenth century many crystalline samples of various cobalt ammonates were synthesized. These compounds are highly colored, and the names given to them (for example, roseo "red", luteo "deep yellow", and purpureo "purple" cobalt chlorides) reflected these colors. After the synthesis of these cobalt ammonates, a great deal of experimental work was performed by various chemists to try to determine what structure these beautiful compounds adopted.

Dmitri Mendeleev published his first periodic table in **1869**.

In the 2nd half of the nineteenth century other ammonates of chromium and platinum were prepared. No theoretical basis was developed to account for the bonding and structure of these compounds.

In *1869-71 Christian Wilhelm Blomstrand* first proposed his *chain theory* to describe the structure of the cobalt ammonate chlorides and other series of transition metal ammonates. It was based upon a completely logical but incorrect analogy with the bonding in organic amines.

Blomstrand, who knew that the fixed valence (*oxidation state*) of cobalt was established at 3, came up with a structure where he linked together cobalt atoms, ammonia groups and chlorine atoms to produce a linear chain structure of $CoCl_3 \cdot 6NH_3$. Since it was based on the prevailing ideas of the time, this was considered to be a reasonable structure.

It was assumed that a metallic atom or ion could replace the hydrogen atoms of ammonia just as organic moieties do in the formation of amines such as methyl amine [CH_3NH_2], dimethyl amine [$(CH_3)_2NH$], and trimethyl amine [$(CH_3)_3N$]. For example, the existence of two forms of $PtCl_4(NH_3)_2$ was rationalized by the two formulas:

$$Cl_3Pt—NH_3—NH_3Cl \quad \text{and} \quad Cl_2Pt(NH_3Cl)_2.$$

Svante Arrhenius was a Swedish chemist best known for his theory that electrolytes, (which are substances that dissolve in water to yield a solution that conducts electricity) are separated or dissociated into electrically charged particles, or ions, even when there is no current flowing through the solution. This resulted in his thesis (**1884**) "*Recherches sur la conductibilité galvanique des électrolytes*" (Investigations on the Galvanic Conductivity of Electrolytes). Arrhenius concluded that when electrolytes are dissolved in water they become to varying degrees split or dissociated into electrically opposite positive and negative ions (cations and anions). In 1903 he was awarded the Nobel Prize for Chemistry.

S.M. Jörgensen (1837-1914) extended Blomstrand's chain theory in **1884**. Before Jörgensen began his extensive studies on the synthesis of "complex" metal compounds it was known that the reaction of metal halides and other salts with neutral molecules could give stable compounds. Many of these compounds could easily be formed in aqueous solutions.

The chain theory is of interest mainly because it became a matter of contentious debate between Jörgensen and Werner, and the study of this heated discussion gives insight into the workings of chemistry and science. The contentious debate between the two rivals (Jörgensen was older, Werner was younger) provided the stimulus for further research. As more experimental data became available, the chain theory required increasing modification. Ultimately this patch-work theory was finally discarded, but it had a long period of popularity despite all its flaws. Discussion has been made about the possible damage that was done to science by this strong belief in an incorrect model, but it was an important part of the process of discovery.

The contentious scientific debate is described in the following passages. Jörgensen, who was a student of Blomstrand, proposed some amendments to Blomstrand's depiction of the cobalt ammonates.
Structures of the cobalt ammonate chlorides as depicted by Blomstrand and Jörgensen are shown in Figure 1.

Figure 1. The Blomstrand-Jörgensen Chain Theory
(a) Blomstrand's structure of $CoCl_3 \cdot 6NH_3$
(b) Jörgensen's structures of the other four members of the cobalt ammonate series plus the iridium substituted compound

(a) $CoCl_3 \cdot 6NH_3$

$$Co \begin{cases} NH_3-NH_3-Cl \\ NH_3-NH_3-Cl \\ NH_3-NH_3-Cl \end{cases}$$

(b) Series

(1) $CoCl_3 \cdot 6NH_3$

$$Co \begin{cases} NH_3-Cl \\ NH_3-NH_3-NH_3-NH_3-Cl \\ NH_3-Cl \end{cases}$$

(2) $CoCl_3 \cdot 5NH_3$

$$Co \begin{cases} Cl \\ NH_3-NH_3-NH_3-NH_3-Cl \\ NH_3-Cl \end{cases}$$

(3) $CoCl_3 \cdot 4NH_3$

$$Co \begin{cases} Cl \\ NH_3-NH_3-NH_3-NH_3-Cl \\ Cl \end{cases}$$

(4) $IrCl_3 \cdot 3NH_3$

$$Ir \begin{cases} Cl \\ NH_3-NH_3-NH_3-Cl \\ Cl \end{cases}$$

First, Jörgensen proposed that these complex compounds were monomeric, i.e. they only contained one metal.

In their laboratories, the number of chloride ions precipitated as silver chloride (see Table 1.) was determined by the addition of aqueous silver nitrate, as represented in the following equation:

$$AgNO_3 \ _{(aq)} + Cl^- \ _{(aq)} \rightarrow AgCl \ _{(s)} + NO_3^- \ _{(aq)}$$

Next, to account for the rates at which various chlorides were precipitated he adjusted the distance of these chloride groups away from the cobalt.

Since the first chloride is precipitated much more rapidly by silver nitrate than the other chlorides, it was placed farther down the chain and under less influence of the cobalt atom. Jörgensen assumed that if the chloride was directly attached to the cobalt, then it would not be available to participate in the precipitation reaction with silver nitrate.

The proposed structures for the first three cobalt ammonate chlorides are shown in Figure 1, and it can be seen in the structure of the second compound in the (b) series that one chloride is directly attached to the cobalt and is therefore unavailable to be precipitated by silver nitrate. In the third compound, two chlorides are similarly pictured, and are thus unavailable for precipitation by silver nitrate.

These changes improved the viability of the chain theory, but a number of unanswered questions remained. For example, why are there only 6 ammonia molecules in the compounds and not 8 or 4? Why do different ammonia molecules that are unique depending on their positions in the chain, not react differently? However, the Blomstrand-Jörgensen theory of the cobalt ammonates was accepted anyway.

Given the success of organic chemists in describing the structural units and fixed atomic valences found in carbon-based compounds, it was natural that these ideas be applied to metal ammonates.

The lack of knowledge about the bonding was reflected in the dot used in the formula to connect $CoCl_3$ to the correct number of ammonias (for example, $CoCl_3 \cdot 6NH_3$).

TABLE 1: The Cobalt Ammonate Chlorides
(Data Available to Blomstrand, Jörgensen, and Werner)

Formula	Conductivity	# of Cl^- ions precipitated	Electrolytic Behavior
NaCl		1	1:1 electrolyte
$CaCl_2$		2	1:2 electrolyte
$LaCl_3$		3	1:3 electrolyte
$CoCl_3 \cdot 6NH_3$	High	3	1:3 electrolyte
$CoCl_3 \cdot 5NH_3$	Medium	2	1:2 electrolyte
$CoCl_3 \cdot 4NH_3$	Low	1	1:1 electrolyte
$CoCl_3 \cdot 4NH_3$	Low	1	1:1 electrolyte
$IrCl_3 \cdot 3NH_3$	Zero	0	nonelectrolyte

The $CoCl_3 \bullet 3NH_3$ compound wasn't known at the time. Jörgensen did, however, manage to prepare, after considerable time and effort, the analogous iridium ammonate chloride, but since it was found to have *no ionizable chlorides* then the only conclusion possible was that it must be a neutral compound.

The Blomstrand-Jörgensen chain theory was in deep trouble due to the research efforts of one of its investigators!

Werner and Coordination Theory

In **1891**, **Alfred Werner**, at age twenty-five, published a paper titled "Contribution to the Theory of Affinity and Valence," where he rejected the usual concepts of valence and affinity, (attraction) of atoms for each other.

In **1892**, the following year, Werner proposed **coordination theory**, which formed the basis for a new field of inorganic chemistry. Werner observed the difficulties that inorganic chemists were having in explaining the bonding in metal "complexes", and he was aware that the established ideas of organic chemistry seemed to lead only into more confusion.

The recognition of the true nature of metal "complexes" was set out in his classic work *Neuere An-schauungen auf dem Gebiete der anorganischen Chemie* (Newer Ideas in Inorganic Chemistry) in 1905; he received the Nobel Prize for this work in 1913.

Accounts say that, "According to his own statement, the inspiration in 1892 came to him like a flash. One morning at two o'clock he awoke with a start: the long-sought solution of this problem had lodged in his brain. He arose from his bed and by five o'clock in the afternoon the essential points of the coordination theory were achieved."

In developing his theory Werner made a major departure from prevailing theory.

Werner discarded Blomstrand's and Jorgensen's chain theory in favor of a centralized construction for metal complexes about which other atoms arranged or coordinated themselves around.

The primary valence, or ionizable valence corresponded to what we call today the **oxidation state**; for cobalt, it is the 3^+ state. The secondary valence is more commonly called **the coordination number** (the number of atoms or groups directly attached to an atom). For cobalt it is six.

Thus, the primary valence is the oxidation state, and the secondary valence is the coordination number.

He found that a few coordination numbers are prevalent, namely two, four, six, and eight, with six by far the most common. Werner proposed likely spatial configurations for those coordination numbers and explained the existence of geometrical isomers.

He predicted the existence of optical isomers, which he eventually isolated, for some cobalt compounds of coordination number 6; these were the first known optically active inorganic compounds.

It is interesting to see how Werner used his postulates to explain the properties of the cobalt (III) chloride-ammonia complexes listed in Table 1.

Thus, the compound $CoCl_3 \cdot 6NH_3$ was represented as $[Co(NH_3)_6]Cl_3$ since there are three readily available chloride ions.

In solution there were therefore four ions, $[Co(NH_3)_6]^{3+}$ and the three chloride ions. This is an important distinction!

The conductivity data (see Table 1.) could then be explained because $CoCl_3 \cdot 6NH_3$ was actually

$[Co(NH_3)_6]Cl_3$, and the following dissolution process occurs in aqueous solution:

$$[Co(NH_3)_6]Cl_3 \rightarrow [Co(NH_3)_6]^{3+} + 3\ Cl^- \quad .$$

It is important for you the student to be able to recognize that some atoms, molecules or ions are bound directly to the metal atom in a "covalent" type fashion, which will be discussed later, and that counter-ions may be present as well which are ionically bonded to the metal complex, just as Na^+ and Cl^- are bonded together in common table salt. In the case of the $[Co(NH_3)_6]Cl_3$ complex, the formula indicates that the ammonia molecules are bound to the cobalt atom *covalently* for lack of a better term at this point, and that the three chloride ions are just counter-ions.

The formulas for the other cobalt ammonate compounds can be determined in a similar way. Two compounds, a violet and a green form of $CoCl_3 \cdot 4NH_3$, were puzzling to chemists of the time because they have the same formula although they are obviously different chemical compounds.

Werner concluded that these two compounds were geometrical isomers and were the *cis* and *trans* forms, where coordinated atoms are at the corners of an octahedron.

As might be expected, Werner's radical departure from long-accepted chemistry theories and bonding concepts met with disapproval and controversy.

Jörgensen criticized Werner's theory on several points: (1) the postulate that all coordinating groups occupy equivalent positions in the coordination sphere, (2) Werner's designation of geometrical isomers of such compounds as $[CoCl_2(NH_3)_4]Cl$, and (3) the prediction of metal compounds that were then unknown.

These criticisms, and others, caused Werner and his associates to diligently produce experimental evidence to support his coordination theory in detail.

The coordination theory that Werner proposed has not been discredited by further research over time, but has served as a framework and a guide for all coordination chemists.

Werner next turned to the geometry of the secondary valence (or coordination number). As shown in Table 2, six ammonia molecules placed about a central metal atom or ion might adopt one of several different common geometries, including hexagonal planar, ***trigonal prismatic***, and ***octahedral***. The table compares some information about the predicted and actual number of isomers for a variety of substituted coordination compounds, in particular the cobalt ammonates.

There is a need to make a few comments about the information in this table before discussing the data. Note that the symbols for the compounds use M for the central metal atom and A's and B's for the ligands. ***Isomers* are defined here as compounds that have the same numbers and types of chemical bonds but differ in the spatial arrangements of those bonds**. The number of predicted isomers refers to the number of arrangements in space that are theoretically possible for the given geometry.

For the planar MA_5B case, for example, there is only one geometry possible, even though there are numerous ways to draw it. Draw these yourself, and make models!!

Table 2. The Number of Actual versus Predicted Isomers of Three Possible Geometries of Coordination Number 6

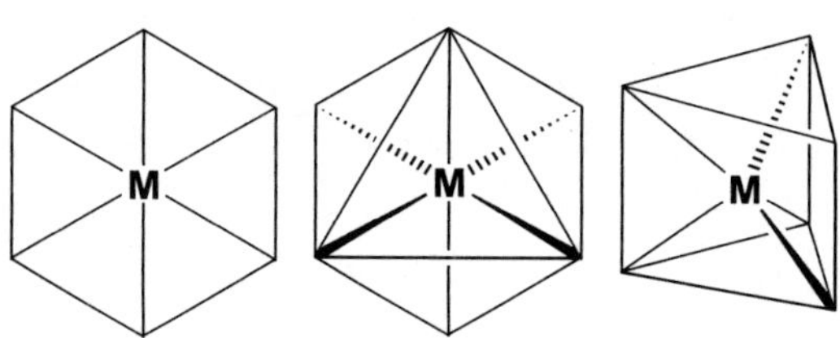

Formula	Planar	Octahedral	Trigonal Prism	Observed
MA_5B	1	1	1	1
MA_4B_2	3	2	3	2
MA_3B_3	3	2	3	2

In comparing different isomers of the same geometry it does not matter which position is occupied by the one B ligand since all the positions are equivalent; thus, there is only one isomer for the MA_5B case.

Now, one can inspect the data in Table 2. In the MA_5B case, only one isomer of the cobalt ammonates could actually be synthesized. This result is consistent with all three of the proposed geometries, and doesn't help in determining which geometry is favored by the cobalt ammonate coordination compounds.

However, for the MA_4B_2 case, Werner could prepare only *two* isomers! Importantly, this actual number of isomers synthesized matched the possible number of isomers predicted for the octahedral geometry. Also, and just as importantly, for the hexagonal planar and trigonal prism cases there were *three* possible predicted isomers for each of those geometries. Assuming that Werner hadn't missed the preparation of an isomer, this indicated that the coordination geometry for six ligands around cobalt is octahedral.

For the MA_3B_3 case only the octahedral geometrical configuration predicts the same number of isomers that were actually synthesized by Werner. By analyzing a large number of series of coordination compounds of different metals, Werner could predict that two isomers would be found for the $CoCl_3 \cdot 4NH_3$ case. These two isomers proved to be difficult to prepare, but in 1907 Werner was finally successful in their synthesis.

Werner synthesized two isomers of $CoCl_3 \cdot 4NH_3$, one a striking violet and the other a bright green color.

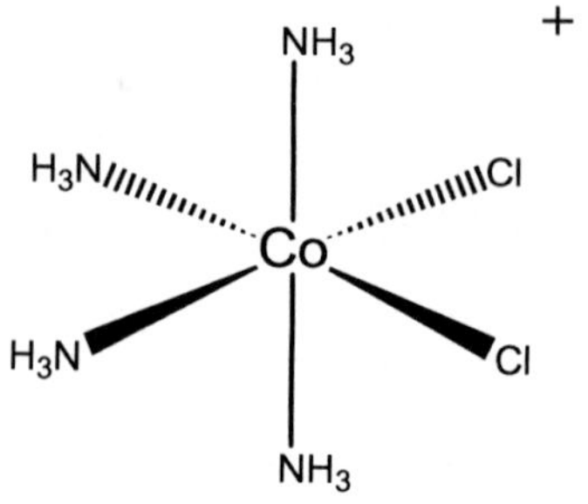

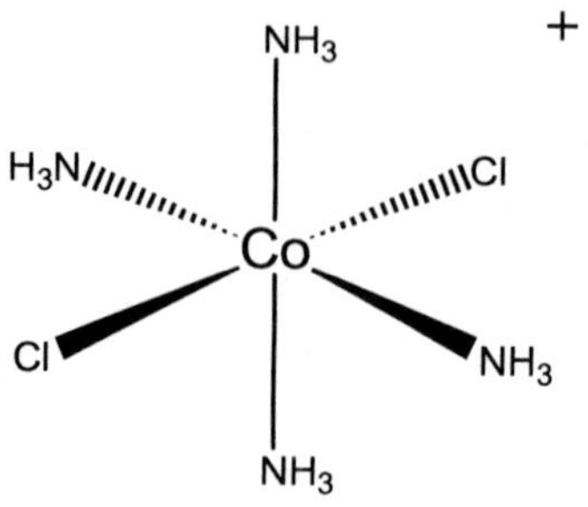

Although all this would not be considered conclusive proof but as "negative" evidence by some philosophers of science---since it was the *absence* of an isomer that constituted the evidence---the case for Werner's coordination theory grew stronger.

However, this negative proof was enough for Jörgensen. In 1907 he dropped his opposition to Werner's "preposterous" coordination theory.

The march of coordination chemistry continued with new discoveries, which forced a rethinking of older organic theories, and the emergence of newer bonding theories.

Once the electron was discovered in the early part of the twentieth century (J.J. **Thomson**, Robert **Millikan**), then **G.N. Lewis** was able to explain some aspects of bonding on the basis of his electron-dot formulas and the octet rule, which we will cover in-depth in Chapter 2.

G. N. Lewis was trying to explain valence to his students in **1902** when he depicted atoms as a concentric series of cubes with electrons at each corner. This "cubic atom" explained the eight groups in the periodic table and presented his theory that chemical bonds are formed by electron sharing to give each atom a complete set of eight, an octet. These early *Lewis Electron Dot Structures* would later be modified in another classroom discussion.

In **1923 Lewis** redefined acids as any atom or molecule with an incomplete "octet" that were thus capable of accepting electrons from another atom; bases were, of course, electron donors. Thus began the useful *Lewis Acid/Base Theory*.

In his book, "The Electronic Theory of Valency", which was published in **1927** as a culmination of many years' interest in the nature of covalent and dative bonds, **Nevil Sidgwick** discussed the electronic interpretation of coordination chemistry and discussed the covalent nature of the chemical bond between metal and ligand, since Werner had shown that it wasn't ionic.

In **1929, Hans Bethe** used J. Becquerel's basic idea (1928) of a crystal field and formulated an exact theory he called *"Crystal Field Theory"* which used symmetry concepts to describe the electronic levels of gaseous metal ions.

In **1931-32, Linus Pauling** used the ideas of Sidgwick and Lewis to discuss the covalent bonding between a metal and a ligand in quantum mechanical language, and used metal orbitals to construct a hypothetical "hybrid orbital" which the metal could use to bond to a ligand in a covalent fashion. This hybrid orbital was described as a d^2sp^3 hybrid orbital. Linus Pauling's valence-bond theory will be discussed in Chapter 2 and Chapter 5.

J.H. Van Vleck made the first real application of Bethe's Crystal Field Theory to the chemistry of the transition metals in **1932**. He succeeded in explaining why the paramagnetism of some of the complexes of the first transition metal series existed using the Crystal Field Theory.

The valence-shell electron-pair repulsion (VSEPR), valence-bond (VB), and molecular orbital (MO) theories were fleshed out in the 1930s. We will cover VSEPR theory (by Sidgwick) in Chapter 2, and discuss valence-bond theory as set out by Linus Pauling. Molecular orbital theory will be discussed later in the text, in Chapter 6.

Crystal Field Theory was largely ignored before WW II, and then utilized heavily in the 1950's once it was "discovered" in the literature.

This marks the beginning of the modern age of coordination chemistry.

Older formulas vs. Modern formulas

The formulation of metal complexes can now no longer be expressed in the older form such as listed in Table 1 since that form does not reveal any structural information. Since two different metal complexes can have the same formula ($CoCl_3 \cdot 4NH_3$ as shown above), then it is necessary to formulate them differently. The formulations must be systematic and describe covalent bonding and the ionic nature of the metal complex as accurately as possible.

This is taken up in the next section.

Anatomy of a metal complex (Modern Formulas)

A transition metal complex has several parts as can be seen in the following scheme: (1) a central metal atom, (2) ligands, and (3) it may possess a counter ion such as an anion.

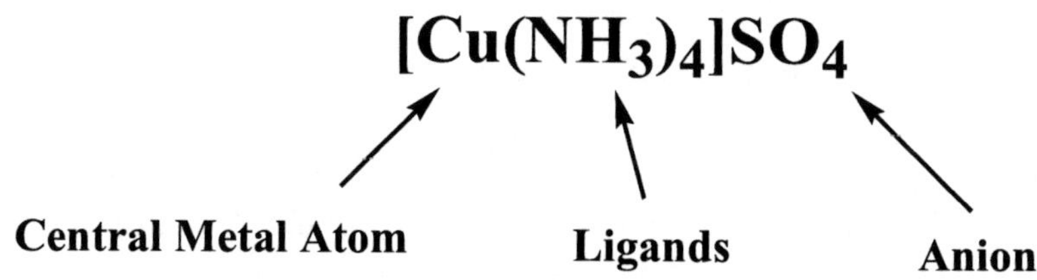

The ligands are bound directly to the central metal atom, and are said to be in the primary coordination sphere. The counter-ions, such as the sulfate anion in the above copper complex, are bound to the metal complex by electrostatic attraction, and are said to be in the secondary coordination sphere.

The above metal complex may dissolve in water, like a normal salt, to produce a cation and an anion:

$$[Cu(NH_3)_4]SO_4 \xrightarrow{\text{water}} [Cu(NH_3)_4]^{2+} + SO_4^{2-}$$

Alternatively, and more rarely, a metal complex may actually be an anion instead of a cation.

An example is shown below:

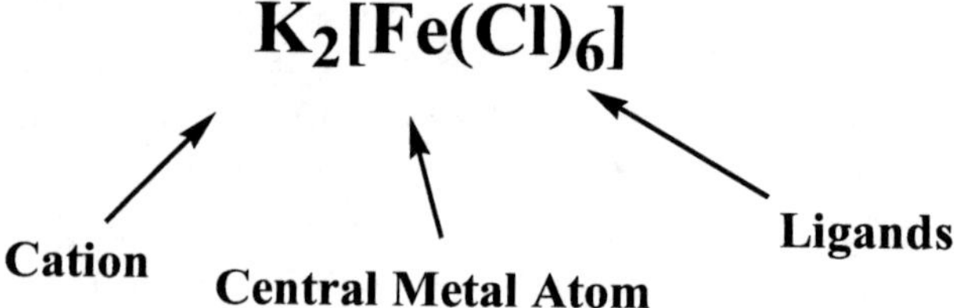

In this case, the iron complex is an anion, and it needs two potassium cations to balance its charge. It may dissolve in water to produce:

$$K_2[Fe(Cl)_6] \xrightarrow{\text{water}} 2\,K^+ + [Fe(Cl)_6]^{2-}$$

Periodic table and trends

The periodic table is divided into three types of elements: (1) *metals*, (2) *semimetals* or *metalloids*, and (3) *nonmetals*. The metals make up the bulk of the periodic table and the chemistry of the transition metals is the focus of this text. The nonmetals are also of importance to the chemistry of the transition metals, since they provide the ligands that react with the metals.

Metals are distinguished chemically from the nonmetals by the absence of negative oxidation states. (There are exceptions in organometallic compounds) Metals have the ability to form positive ions, ionic bonds, and basic oxides.

The vertical columns or *Groups* of the periodic table contain elements having similar chemical and physical properties, and several groups of elements have distinct names that you must know.

Elements in Group I are known as the *alkali metals* (except for hydrogen). The word alkali comes from the ancient Arabic word *al-qali*, which is the word for plants whose ashes give water solutions that were slippery and burned the skin. The alkali metals are always found as +1 charged cations.

Similarly, the Group II metals are called the **alkaline earth metals**, since the metal hydroxides form basic solutions and the Group II cations, which are always +2 charged, usually form insoluble salts.

The two rows at the bottom of the periodic table, the so-called **f-block elements**, are the **lanthanides** and the **actinides**, which are also called the **inner transition metals**. The f-block elements have a common +3 oxidation state.

Group VII elements are called the **halogens**. Group VI elements are called the **chalcogens** from the Greek word, *khalkos*, for copper, whose ores often contain them. These are nonmetals.

The **transition metals** are the so-called **d-block elements** (which follow Group II elements) that fill the fourth, fifth, and sixth periods in the center of the periodic table. The transition metals have varying oxidation states, with many different oxidation states each, with the most common oxidation states being +2, and +3. Transition metals can use electrons from more than one subshell, so they exhibit multiple oxidation states. As a general rule, the higher the oxidation state of the metal, the more covalent is the bonding.

Chemists often use the term *transition metal* for elements that have partially filled d-orbitals. Thus, Zn, Cd, and Hg are often excluded from the list, and it is no surprise that they do exhibit chemistry that resembles the representative main group elements.

Transition metals and their compounds or complexes have characteristic properties such as: (a) *variable oxidation states*, (b) *small, compact atoms*, which decrease across a period and then increases towards the end (c) *stable compounds* of varying covalency in far more numerous examples, geometries and types than do the metals from the main group, (d) *magnetic properties* such as paramagnetism and diamagnetism, (e) *highly colored complexes* that span the rainbow (names of some of the transition metals express this fact---iridium, rhodium, chromium, etc.), (f) *catalytic activity* such as with enzymes and industrially used catalysts and (g) *variable coordination geometries* such as square planar complexes, which are unknown for the main group metals. It is the study of these fascinating properties of the transition metals and their coordination chemistry that is the focus of this book.

Electronic Configurations

On pages 26 and 27 are listed the ***electronic configurations*** of the first 86 elements, and the atomic radii of selected elements, which include the first three rows of the transition metals. The transition metals are listed in bold.

Electron configurations are extremely important to inorganic chemists and help us to understand the reactivity of transition metal ions. The loss of certain numbers of electrons from metal atoms to form metal ions can be understood much better once the electron configuration information is available.

For example, scandium is nearly always found in the Sc^{+3} oxidation state. This makes sense because the electron configuration of Sc metal is: $Sc = [Ar] 4s^2 3d^1$. If Sc metal loses three electrons then the electron configuration of Sc^{+3} would be that of Ar.

Since most of the transition metals have ns^2 (where n=3, 4 or 5) as part of their electron configuration, then it is no surprise that the +2 oxidation state for transition metals is so very common. In transition metal complexes it is assumed by convention that the ns^2 electrons are always the first to go when the metal is oxidized. Also, as it will be restated in the section on electron counting in Chapter 7., the electron configuration given here is only for isolated atoms of that element (gaseous atoms). If ligands are around a metal in some metal complex, then the electronic configurations noted here are not considered to be valid.

Atomic Radii

Radii of the transition metals are determined in the solid state by using X-Ray diffraction to measure the metal-metal distances. The radii of the transition metals vary over a narrow range in each period, which is determined by the outermost valence electrons, which are in the 4s, 5s or 6s orbitals for each of the three periods respectively.

A graph of atomic radii of the 3d, 4d and 5d transition metal series.

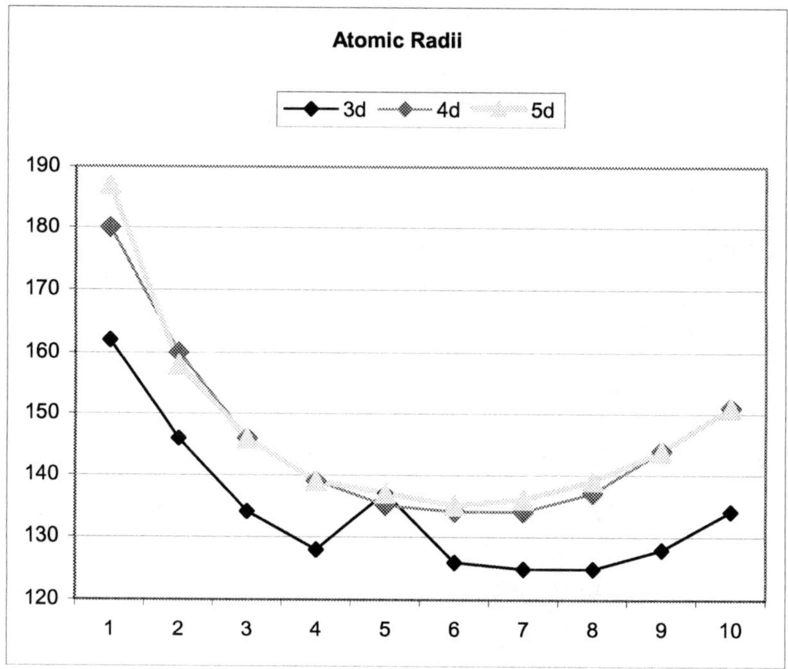

An interesting observation is that the radii of the 2nd and 3rd row transition metals (the fifth and sixth period) are almost identical. The reason for this is that the inner transition elements, the lanthanide elements, are inserted into the Periodic Table between La and Hf. The filling of the 4f orbitals is accompanied by a steady decrease in the atomic radii to the point that the 3rd row transition elements have acquired a size almost identical with those in the 2nd row. This significant effect is called the ***lanthanide contraction***.

Electron Configurations of the First 86 Elements

Z	Element	Electron Configuration
1	H	$1s^1$
2	He	$1s^2$

Z	Element	Electron Configuration
3	Li	[He] $2s^1$
4	Be	[He] $2s^2$
5	B	[He] $2s^2 2p^1$
6	C	[He] $2s^2 2p^2$
7	N	[He] $2s^2 2p^3$
8	O	[He] $2s^2 2p^4$
9	F	[He] $2s^2 2p^5$
10	Ne	[He] $2s^2 2p^6$

Z	Element	Electron Configuration
11	Na	[Ne] $3s^1$
12	Mg	[Ne] $3s^2$
13	Al	[Ne] $3s^2 3p^1$
14	Si	[Ne] $3s^2 3p^2$
15	P	[Ne] $3s^2 3p^3$
16	S	[Ne] $3s^2 3p^4$
17	Cl	[Ne] $3s^2 3p^5$
18	Ar	[Ne] $3s^2 3p^6$

Z	Element	Electron Configuration	Atomic Radii
19	K	[Ar] $4s^1$	230
20	Ca	[Ar] $4s^2$	197
21	Sc	[Ar] $4s^2 3d^1$	162
22	Ti	[Ar] $4s^2 3d^2$	146
23	V	[Ar] $4s^2 3d^3$	134
24	Cr	[Ar] $4s^1 3d^5$	128
25	Mn	[Ar] $4s^2 3d^5$	137
26	Fe	[Ar] $4s^2 3d^6$	126
27	Co	[Ar] $4s^2 3d^7$	125
28	Ni	[Ar] $4s^2 3d^8$	125
29	Cu	[Ar] $4s^1 3d^{10}$	128
30	Zn	[Ar] $4s^2 3d^{10}$	134
31	Ga	[Ar] $4s^2 3d^{10} 4p^1$	135
32	Ge	[Ar] $4s^2 3d^{10} 4p^2$	
33	As	[Ar] $4s^2 3d^{10} 4p^3$	
34	Se	[Ar] $4s^2 3d^{10} 4p^4$	
35	Br	[Ar] $4s^2 3d^{10} 4p^5$	
36	Kr	[Ar] $4s^2 3d^{10} 4p^6$	

Z	Element	Electron Configuration	Atomic Radii
37	Rb	[Kr] $5s^1$	247
38	Sr	[Kr] $5s^2$	215
39	Y	[Kr] $5s^2 4d^1$	180

Z	Element	Electron Configuration	Atomic Radii
40	**Zr**	**[Kr] $5s^2\,4d^2$**	**160**
41	**Nb**	**[Kr] $5s^1\,4d^4$**	**146**
42	**Mo**	**[Kr] $5s^1\,4d^5$**	**139**
43	**Tc**	**[Kr] $5s^1\,4d^6$**	**135**
44	**Ru**	**[Kr] $5s^1\,4d^7$**	**134**
45	**Rh**	**[Kr] $5s^1\,4d^8$**	**134**
46	**Pd**	**[Kr] $5s^0\,4d^{10}$**	**137**
47	**Ag**	**[Kr] $5s^1\,4d^{10}$**	**144**
48	**Cd**	**[Kr] $5s^2\,4d^{10}$**	**151**
49	In	[Kr] $5s^2\,4d^{10}\,5p^1$	167
50	Sn	[Kr] $5s^2\,4d^{10}\,5p^2$	154
51	Sb	[Kr] $5s^2\,4d^{10}\,5p^3$	
52	Te	[Kr] $5s^2\,4d^{10}\,5p^4$	
53	I	[Kr] $5s^2\,4d^{10}\,5p^5$	
54	Xe	[Kr] $5s^2\,4d^{10}\,5p^6$	

Z	Element	Electron Configuration	Atomic Radii
55	Cs	[Xe] $6s^1$	267
56	Ba	[Xe] $6s^2$	222
57	**La**	**[Xe] $6s^2\,5d^1$**	**187**
58	Ce	[Xe] $6s^2\,5d^0\,4f^2$	182
59	Pr	[Xe] $6s^2\,5d^0\,4f^3$	182
60	Nd	[Xe] $6s^2\,5d^0\,4f^4$	182
61	Pm	[Xe] $6s^2\,5d^0\,4f^5$	181
62	Sm	[Xe] $6s^2\,5d^0\,4f^6$	180
63	Eu	[Xe] $6s^2\,5d^0\,4f^7$	204
64	Gd	[Xe] $6s^2\,5d^0\,4f^8$	179
65	Tb	[Xe] $6s^2\,5d^0\,4f^9$	178
66	Dy	[Xe] $6s^2\,5d^0\,4f^{10}$	177
67	Ho	[Xe] $6s^2\,5d^0\,4f^{11}$	176
68	Er	[Xe] $6s^2\,5d^0\,4f^{12}$	175
69	Tm	[Xe] $6s^2\,5d^0\,4f^{13}$	174
70	Yb	[Xe] $6s^2\,5d^0\,4f^{14}$	193
71	Lu	[Xe] $6s^2\,5d^1\,4f^{14}$	174
72	**Hf**	**[Xe] $6s^2\,5d^2\,4f^{14}$**	**158**
73	**Ta**	**[Xe] $6s^2\,5d^3\,4f^{14}$**	**146**
74	**W**	**[Xe] $6s^2\,5d^4\,4f^{14}$**	**139**
75	**Re**	**[Xe] $6s^2\,5d^5\,4f^{14}$**	**137**
76	**Os**	**[Xe] $6s^2\,5d^6\,4f^{14}$**	**135**
77	**Ir**	**[Xe] $6s^2\,5d^7\,4f^{14}$**	**136**
78	**Pt**	**[Xe] $6s^1\,5d^9\,4f^{14}$**	**139**
79	**Au**	**[Xe] $6s^1\,5d^{10}\,4f^{14}$**	**144**
80	**Hg**	**[Xe] $6s^2\,5d^{10}\,4f^{14}$**	**151**
81	Tl	[Xe] $6s^2\,5d^{10}\,4f^{14}\,6p^1$	171
82	Pb	[Xe] $6s^2\,5d^{10}\,4f^{14}\,6p^2$	175
83	Bi	[Xe] $6s^2\,5d^{10}\,4f^{14}\,6p^3$	
84	Po	[Xe] $6s^2\,5d^{10}\,4f^{14}\,6p^4$	
85	At	[Xe] $6s^2\,5d^{10}\,4f^{14}\,6p^5$	
86	Rn	[Xe] $6s^2\,5d^{10}\,4f^{14}\,6p^6$	

Terms and Definitions Chapter 1

Alfred Werner ---
Christian Blomstrand ---
S.M. Jörgensen ---
Chain Theory ---

Ionizable Chlorides ---

Oxidation State ---

Coordination Number ---

Isomers ---

Primary and Secondary Valence ---

Metals ---
Semimetals or *Metalloids* ---
Nonmetals ---

Groups ---
Alkali Metals ---
Alkaline Earth Metals ---
Halogens ---
Chalcogens ---

Transition Metals ---
Inner Transition Metals ---
d-Block Elements ---
f-Block Elements ---
Lanthanides ---
lanthanide contraction ---
Actinides ---
Electronic Configurations ---

Chapter 1 Problems and Exercises

1. What geometries are associated with coordination numbers 2, 4 and 6?

2. If one mole of the metal complex $[Co(NH_3)_6]Cl_3$ was dissolved in water, and then excess $AgNO_3$ (silver nitrate) was added to the solution, how many moles of AgCl would precipitate out of solution?   (a) zero (b) 1 (c) 2 (d) 3

3. Conductivity measurements confirm that 1 mole of $CrBr_3 \cdot 4NH_3$ dissociates into 2 moles of ions in aqueous solution.
Write:
   (a) the modern formula of the coordination complex and
   (b) an equation for the dissociation.

4. For the complex $[Fe(H_2O)_6]^{3+}$, identify
   (a) the coordination number of iron
   (b) the coordination geometry of iron
   (c) the oxidation state of the iron

5. For the complex $[Mn(NH_3)_6]^{3+}$, identify
   (a) the coordination number of manganese
   (b) the coordination geometry of manganese
   (c) the oxidation state of manganese

6. For the complex $[Co(NH_3)_5Cl]^{2+}$, identify
   (a) the coordination number of cobalt
   (b) the coordination geometry of cobalt
   (c) the oxidation state of cobalt

7. For the complex $[Co(NH_3)_4Cl_2]^{+}$, identify
   (a) the coordination number of cobalt
   (b) the coordination geometry of cobalt
   (c) the oxidation state of cobalt

8. Combinations of cobalt(III), ammonia (NH_3), nitrite (NO_2^-) anions, and potassium (K^+) cations result in the formation of a series of seven coordination compounds.

(a) Write the modern formulas for the members of this series. (*Hint*: Not all seven compounds contain all four of the components.)
(b) How many ionic nitrites would there be in each compound?
(c) How many isomers would each compound have, assuming that each has an octahedral coordination sphere?

9. Combinations of iron(II), H_2O, Cl^-, and NH_4^+ can result in the formation or a series of seven coordination compounds, one of which is $[Fe(H_2O)_6]Cl_2$.

(a) Write the modern formulas for the other members of the series. (*Hint*: Not all seven compounds contain all four of the components.)
(b) How many chlorides could be precipitated from each compound by reaction with aqueous silver nitrate?
(c) How many isomers would each compound have, assuming that each has an octahedral coordination sphere?

10. Combinations of platinum(II)cations, ammonia molecules, thiocyanate anions, and ammonium cations result in the formation of a series of five coordination compounds. Write the modern formulas for the members of this series. (*Hint*: Not all five compounds contain each of the components.)

11. Combinations of palladium(II) cations, triphenylphosphine molecules, chloride anions, and ammonium cations result in the formation of a series of five coordination compounds.

(a) Write the modern formulas for the members of this series. (*Hint*: Not all five compounds contain each of the components.)

(b) Which compound or compounds would have the greatest conductivity in solution? Briefly explain your answer.

(c) Write the formula for the neutral member of this series as Werner would have written it. Assuming that palladium(II) has a square planar coordination sphere, how many isomers would this compound have?

12. If dissolved in water, how many ions would be formed from dissolution of octahedral $[Co(NH_3)_5Cl_3]$?
    What are the ions?

$$[Co(NH_3)_5Cl_3] \quad \rightarrow$$

13. If dissolved in water, how many ions would be formed from dissolution of octahedral $[K_4FeCl_6]$?
    What are the ions?

$$[K_4FeCl_6] \quad \rightarrow$$

14. How many moles of Br- ion would be produced by dissolution of 1.5 moles of $[Co(NH_3)_5Br]Br_2$ in water?

15. If 1.96 grams of $[Co(NH_3)_6]Cl_3$ was reacted with an excess of silver nitrate, how much silver chloride (AgCl) would precipitate out of solution?

16. If 2.54 grams of $Ba_3[FeCl_6]_2$ reacted with sodium sulfate (Na_2SO_4) how much barium sulfate $(BaSO_4)$ would precipitate out of solution?

17. What would be the concentration (in molarity) of bromide ion in solution formed from 5.0 grams of $[Co(NH_3)_4Br_2]Br$ in 100ml of solution?

18. The octahedral rhodium (III) metal complex with the overall formula $[Rh(NH_3)_5Cl_3]$ was dissolved in water. What were the ions formed in aqueous solution?

$$[Rh(NH_3)_5Cl_3] \quad \rightarrow$$

19. The metal complex $(NH_4)_3[FeCl_6]$ was dissolved in water.
What were the ions formed in aqueous solution?

$$(NH_4)_3[FeCl_6] \quad \rightarrow$$

20. Coordination compounds of formula MA_4 might be <u>square planar</u> or <u>tetrahedral</u>. How many isomers would you predict to exist for compounds of formula MA_2B_2 for these two geometries? $Pt(NH_3)_2Cl_2$ has two known isomers, and $[CoBr_2I_2]^-$ has only one.
    Speculate on the possible structures of these complexes by drawing them.

21. Using only a periodic table to locate the element, write the electron configuration for
(a) Os   (b) Co (c) Ni (d) Ru (e) Cu (f) Re

22. What is the highest possible oxidation state for each of the following transition metals?
(a) Zr (b) Ta (c) Mn (d) Nb (e) Tc (f) Y

23. Identify two transition metal ions with each of the following electron configurations:
    (a) $[Ar]\ 3d^6$
    (b) $[Ar]\ 3d^5$
    (c) $[Ar]\ 3d^{10}$
    (d) $[Ar]\ 3d^8$

24. Give the electron configuration for each of the ions listed.

(a) Pt^{2+}
(b) Ni^{2+}
(c) Co^{3+}
(d) Ir^{3+}
(e) Ti^{4+}
(f) Zr^{4+}
(g) Cu^{2+}
(h) Ag^{+}
(i) Cr^{3+}
(j) Mo^{3+}
(k) Mn^{3+}
(l) Tc^{3+}

25. What is the coordination number of ruthenium in $K_4[Ru(CN)_6]$?

26. What is the oxidation number of iridium in $[IrCl(NH_3)_5]Cl_2$?

27. In the d-block elements, the third-row metallic radii are about the same as the second-row radii because of ...
    (a) greater shielding of f-electrons.
    (b) greater shielding of d-electrons.
    (c) the lanthanide contraction.
    (c) the increased number of d-electrons.

28. Draw the structures for the isomers that are predicted for the three different geometries (hexagonal planar, octahedral, and trigonal prismatic) that are listed in Table 2.

Advanced (see Chapter 3. for bidentate ligands)

29. Jörgensen synthesized $[CoCl_2(en)_2]Cl$ that came in two forms named for their colors: violeo and praseo. Werner cited the existence of these two (and only two) isomers as proof of an octahedral coordination sphere.

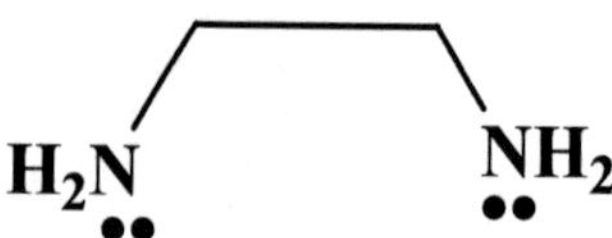

(a) Recalling that en stands for the bidentate ethylenediamine molecule, draw structural formulas for each isomer. (*Hint*: The ethylenediamine molecule can span only adjacent positions in the octahedron.)

(b) Suppose the coordination sphere for coordination number 6 had been trigonal prismatic instead of Werner's octahedral. How many isomers would this compound have had under that assumption? Draw structural formulas for each. (*Hint*: The ethylenediamine molecule can span only adjacent positions on the triangular and rectangular sides of the trigonal prism.)

(c) Suppose the coordination sphere for coordination number 6 had been hexagonal planar instead of Werner's octahedral. How many isomers would this compound have had under that assumption? Draw structural formulas for each. (*Hint*: The ethylenediamine molecule can span only adjacent positions on the hexagon.)

Chapter 2. Lewis Electron Dot Structures and Valence Shell Electron Pair Repulsion Theory

I. Lewis Electron Dot Structures

Lewis Electron Dot Structures were designed by ***G.N. Lewis (1875-1946)*** to show the comparison of the bonding power of an atom with the number of available (valence shell) electrons.

For example, carbon normally forms four bonds, and it has 4 available, or valence shell, electrons. Thus, it is drawn with 4 available electrons, whereas nitrogen has five available electrons but usually forms only three bonds. Most elements acquire an ***octet*** set of eight valence electrons (four pairs) in forming chemical bonds. The first and second row elements can be drawn with their representative Lewis Electron Dot Structures:

$$\dot{H} \qquad \ddot{He}$$

$$\dot{Li} \qquad \dot{Be}\cdot$$

$$\dot{B}\cdot \qquad \cdot\dot{C}\cdot \qquad \cdot\ddot{N}\cdot \qquad \cdot\ddot{O}: \qquad \cdot\ddot{F}: \qquad :\ddot{Ne}:$$

The reaction of two chlorine atoms to form molecular chlorine gas and also the reaction to show the production of the ionic compound sodium chloride is shown using Lewis electron dot structures:

$$:\ddot{Cl}\cdot \quad + \quad \cdot\ddot{Cl}: \quad \longrightarrow \quad :\ddot{Cl}-\ddot{Cl}:$$

$$Na \quad + \quad \cdot\ddot{Cl}: \quad \longrightarrow \quad Na^+ \quad :\ddot{Cl}:^-$$

In the first reaction shown above both Cl atoms acquire an "octet" of valence electrons and are thus "satisfied". In the second reaction, sodium loses its only valence electron, and Cl gains an electron so that both acquire a noble gas configuration (thus effectively acquiring an octet).

Lewis structures are used to show the molecular structures for simple molecules, both organic and inorganic, and are primarily used to describe compounds in organic chemistry and their reactions. Since ionic compounds aren't described as usefully with Lewis electron dot structures, then ionic compounds usually aren't depicted in this manner.

VSEPR theory also utilizes Lewis electron dot structures to describe the simple molecular geometry in small organic and inorganic molecules.

The rules used to draw Lewis electron dot structures are given below:

(1) Sum up the total number of valence electrons for all the atoms in the molecule. For a polyatomic anion, add the number of negative charges to the total, and for a polyatomic cation subtract the number of positive charges from the total.

(2) Draw the skeletal structure for the molecule by connecting bonded pairs of atoms by a bond dash. Each ***bonding pair of electrons*** is denoted by a dash.

(3) Place electrons around the outer atoms so that each atom has an octet. ***(The octet rule)*** Hydrogen can only have a duet of electrons, and that must be its bonding pair.

(4) The remaining electrons (subtract the number assigned so far from the original total from step 1) are assigned in pairs to the central atom as ***lone pairs of electrons***.

(5) A multiple bond is likely if the central atom has fewer that an octet after step 4. A pair of electrons from an outer (peripheral) atom can be used to form a ***multiple bond***.

Rules of thumb:
(1) Central atoms are often written first in chemical formulas.
(2) Choose the atom with the lowest ionization energy for the central atom.
(3) Arrange atoms symmetrically around the central atom. (SO_2 is O-S-O, not S-O-O)

Resonance Structures

More than one valid Lewis structure may be drawn for some molecules that have multiple bonds, and these are called **resonance structures**. An example is the nitrate ion $(NO_3)^-$. The following scheme shows the three possible resonance structures that can be drawn for the nitrate ion.

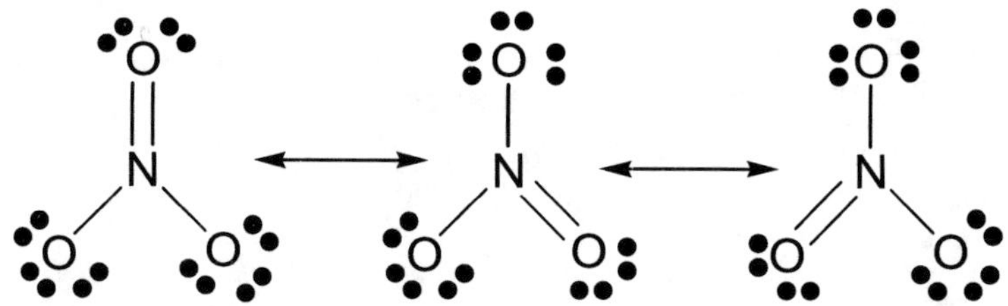

In the case of multiple bonded compounds such as nitrate, Lewis structures actually aren't a valid description of the molecule since there are no separate single and double bonds in the nitrate ion. ***The x-ray crystal structure of the nitrate ion shows that all the N-O bond lengths are exactly the same length.*** The pair of electrons that makes up the double bond between one of the N-O bonds is actually ***delocalized*** throughout the whole molecule. Thus, the bond order between the N and O atoms is actually one and 1/3, which points out a failure in the use of Lewis structures.

Formal Charge
Formal Charge =
Valence e⁻s – (Lone Pair electrons + ½ of the number of electrons in bonds)

It may be hard to tell which atom is the central atom for certain linear molecules such as CO_2, and N_2O. Use of formal charge helps determine which structure is more likely. Structures with the lowest formal charges are likely to have the lowest energy, and therefore be the most stable. Also, the molecule is most stable when any negative charge resides on the most electronegative atom.

For example, the correct structure of CO_2 can be determined by calculating the formal charge for two likely structures. The formal charges are listed above the atoms:

$$\overset{0}{O}=\overset{0}{C}=\overset{0}{O} \qquad \overset{0}{:\ddot{O}}=\overset{+2}{O}=\overset{-2}{C:}$$

As can be seen, the structure with the carbon atom in the center has the lowest formal charges, and therefore the more stable structure. The formal charges for two possible structures for N_2O are shown below:

$$\overset{-1}{:\ddot{N}}=\overset{+1}{N}=\overset{0}{O:} \qquad \overset{-1}{:\ddot{N}}=\overset{+2}{O}=\overset{-1}{N:}$$

Of the structures shown above for N_2O, which do you expect to be the correct structure? (That's right, the 1st one!)

Hypervalent compounds (Expanded Valence Shells)

Some molecules have central atoms that violate the octet rule and have more than four electron pairs around the central atom.

For example:

$$P_4 + 6\,Cl_2 \rightarrow 4\,PCl_3 \qquad PCl_3 + Cl_2 \rightarrow [PCl_4]^+ \, [PCl_6]^-$$

$$\text{and} \quad [PCl_4]^+ \, [PCl_6]^- \rightarrow 2\,PCl_5$$

The complex ion $[PCl_6]-$ has a phosphorous atom that violates the octet rule because it has six chlorides around it, each with a pair of bonding electrons for a total of twelve electrons around the central atom, and PCl_5 has ten.

The compound PCl_5, and its fluorine analogue PF_5 exhibit a trigonal bipyramid geometry which we will discuss shortly.

Nitrogen, which is a group V element like phosphorous, cannot form the analogous compounds NCl_6 or NF_6.

Why?

Even though N and P are in the same group, it must be noticed that N is a 2nd row element, whereas P is a 3rd row element.

The simplest explanation is that phosphorous, since it is a 3rd row element, has available d orbitals that nitrogen does not; therefore using *valence bond theory*, nitrogen can only form sp^3 hybrid orbitals whereas P can form dsp^3 (five bonds) and d^{2}sp^3 (six bonds) hybrid orbitals and expand its octet.

Valence bond theory was put forth by Linus Pauling in his book "The Nature of the Chemical Bond" which was dedicated to G. N. Lewis. The next section briefly discusses the important points of valence bond theory with respect to geometry and molecular shape.

Valence-Bond Theory

Valence bond theory is based on the Lewis electron dot concept that covalent bonding results from sharing of electron pairs between atoms. The valence bond approach is most useful to organic chemists who use the theory very successfully.

The result of using the valence orbitals to form a mixed orbital (usually the s and the three p orbitals) is a *hybrid orbital*. The number of hybrid orbitals formed by this approach equals the number of atomic orbitals that are used to form the hybrid orbitals. Not only s and p orbitals can be used to form hybrid orbitals, but Pauling postulated that d orbitals could be used as well. The formation of hybrid orbitals using valence-bond theory can be used successfully to explain and account for molecular shapes.

Table 2.1
Hybridization of orbitals to explain molecular shape.

Hybridization	# of hybrid orbitals	Molecule geometry
sp	2	Linear
sp^2	3	Trigonal planar
sp^3	4	Tetrahedral
dsp^3	5	Trigonal bipyramidal
d^2sp^3	6	Octahedral

In Chapter 5 we will discuss the shortcomings of valence bond theory in dealing with transition metal complexes, and why it isn't used to explain the bonding and physical properties of these transition metal complexes.

Limitations of VSEPR Theory

The majority of molecules that have a molecular geometry that is trigonal bipyramid or octahedral are hypervalent compounds. An exception to this is the transition metal coordination compounds that we have discussed in Chapter 1., and these transition metal coordination compounds cannot be adequately described by VSEPR theory.

An example of this breakdown of Lewis electron dot structures to explain the geometry of metal complexes can be found in the simplest Werner complexes. For example, the cobalt(III) compound $Co(NH_3)_6^{3+}$ would not have octahedral molecular geometry around the central metal atom. Cobalt (III) has six valence electrons itself, plus twelve more electrons from the six ammonia ligands, giving a total of 18 electrons around the central atom. A total of 18 electrons around the central atom would give a structure that would have nine pairs of electrons around the central atom and would not be octahedral.

Therefore, we will limit the use of Lewis electron dot structures and VSEPR theory to explain the molecular and electronic geometry of only simple molecules and ligands.

Electron Deficient Compounds

Some molecules have central atoms that have less than an octet of electrons. These compounds are called *electron deficient* compounds. Examples of electron deficient compounds are often found with central atoms of the elements boron, beryllium, and aluminum. An example is BF_3, boron trifluoride, which is considered to be a *Lewis acid*. (See Chapter 3. for more discussion on *Lewis Acids* and *Lewis Bases*) Other examples are BeH_2, and trimethyl aluminum.

Examples:

Two valid Lewis structures for the very reactive molecule H_2BF are shown below. One structure shows that boron does not have an octet of electrons, whereas the second structure does.

Calculate the formal charge on each atom in the structures. Which structure is more reasonable in your opinion? Why?

II. Introduction To VSEPR Theory, Electronic Geometry, and Molecular Shapes.

Valence Shell Electron Pair Repulsion Theory (VSEPR) is a practical, though imperfect, means of predicting the basic geometry of many covalent molecules and molecular ions. Many molecules have one atom that is bonded to two or more other atoms. These atoms are called ***central atoms***. Atoms that are bonded to only one other atom are *outer* or ***peripheral atoms***. This is important in two respects, one is that many simple inorganic molecules are best described in this way, and two, the central atom theme compares well to the metal atom geometries found in transition metal coordination compounds.

A central atom will be surrounded by a number **of *regions of high electron density (RHED)***. There are two types of RHED, shared and unshared. An unshared RHED is simply a lone pair of electrons. <u>A shared RHED can be two electrons (a single bond), four electrons (a double bond), six electrons (a triple bond), or a non-integral number of electrons (a resonance bond).</u> For the purposes of VSEPR, all shared RHED are the same. *Most* of the time the RHED is simply an electron pair--hence the name of the theory is "Valence Shell Electron Pair Repulsion Theory", and not "Valence Shell Region of High Electron Density Repulsion Theory".

VSEPR is based on the following principles:

<u>Principle 1</u>: The RHED around a central atom repel one another. Thus, the RHED form angles in which they are as far from one another as possible.

<u>Principle 2</u>: Unshared RHED ("lone pairs") take up a little more room than shared RHED.

From these two principles, one can qualitatively, and sometimes fairly quantitatively, predict bond angles in an impressive variety of molecules and molecular ions.

III.   Lewis Electron Dot Structures

The first step in using VSEPR is to determine the Lewis electron dot structure of the molecule. This will answer two fundamental questions essential to applying VSEPR--how many RHED surround the central atom and how many of these are unshared RHED?

IV.   Applying VSEPR

One can speak of two kinds of geometry that exist simultaneously in a molecule or molecular ion, the electronic geometry and the molecular geometry. *Electronic geometry* is the geometry of all RHED around the central atom, whether shared or unshared. There are only 5 common electronic geometries. These correspond to 2, 3, 4, 5, and 6 RHED around the central atom.

Molecular geometry is the geometry of the atomic centers in the molecule. Thus, the molecular geometry, while it is influenced by the electronic geometry, is determined by considering how many shared RHED and how many unshared RHED are around the central atom.

A. Two Regions of High Electron Density: Linear geometry

Imagine an atom with two RHED around it. Applying Principle 1, these repel each other and try to get as far apart as possible. It is not hard to imagine that this is achieved by forming a bond angle of $180°$, which gives a linear geometry.

The most common example of linear geometry is CO_2. In the electron dot structure, each oxygen is double bonded to the carbon atom, O=C=O. (We'll ignore unshared pairs on the outer atoms--they aren't important to the geometry.) The two RHED around the carbon (double bonds in this case) repel each other and this repulsion causes the molecule to become *linear*. Because both RHED are shared, the molecular geometry is also linear.

Another example is hydrogen cyanide, HCN. The electron dot structure, H:C:::N, has a single bond between the H and the C, and a triple bond between the C and the N (again ignoring unshared pairs on outer atoms, in this case on the nitrogen). Once again, the RHED (a single and a triple bond) repel, and as in CO_2, both the electronic and molecular geometry is "linear".

B. Three Regions of High Electron Density

1. No unshared pairs, **AB_3**
(A = central atom, B = peripheral atom)

Imagine two RHED around a central atom. Now imagine bringing one more RHED in. Because of the mutual repulsion, the two original RHED will bend away from the incoming RHED, forming a triangle of RHED. The triangle lies in a plane, and so the electronic geometry with three RHED is known as *trigonal planar*.

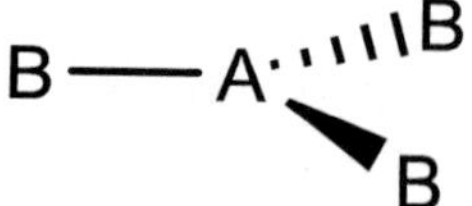

Trigonal Planar

For example, consider the electron deficient compound BF_3. The Lewis dot structure tells us we have three RHED around the B, three single bonds. All the RHED are shared so the molecular geometry is also trigonal planar. All three F and the B lie in the same plane, and the F-B-F bond angles are all 120°.

The same goes for NO_3^-, a molecule that exhibits resonance. In this case there are again three RHED, only this time each resonance structure has one double and two single bonds. What's important, however, is that three RHED are present, and so the geometry is still trigonal planar.

2. One unshared pair, **AB$_2$U**
(A = central atom, B = peripheral atom, U = unshared pair)

Now, suppose one of the RHED is unshared (a lone pair). This occurs in SO_2, where the Lewis structure indicates a lone pair on the sulfur, and a resonance single/double bond between the sulfur and each oxygen atom. Once again, the electronic geometry is trigonal planar, but the molecular geometry, the geometry of the atoms, is **bent**. The angle of bending is *about* 120°, but because the unshared RHED takes a little more room, the bond angle bends to slightly less than 120°, (about 119.5°). Note the difference between SO_2 and CO_2. CO_2 is linear, but SO_2 is bent because of the presence of the lone pair on the sulfur atom.

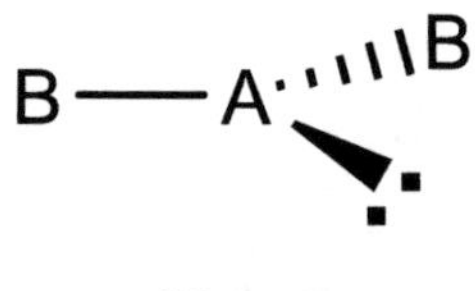

Bent

C. Four Regions of High Electron Density

This is a very common case, because if all four are single bonds, then the octet rule is satisfied for the central atom.

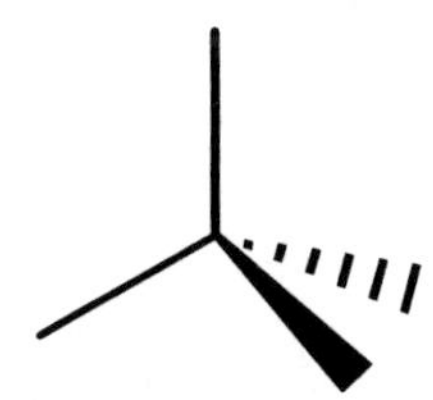

Go back to our trigonal planar situation, with, for example, three electron pairs around a central atom. Now bring in a fourth electron pair perpendicularly. The three originals will be repelled, and will bend out of the plane. The resulting geometry is that of a "tetrahedron". So, the electronic geometry when four RHED are present is **tetrahedral**. The bond angles in the tetrahedron are all ~109.5°.

1. No unshared electron pairs: **AB$_4$**

If there are no unshared RHED, the molecular geometry is also tetrahedral. This is the case for CH_4. The four RHED around the carbon (electron pairs in this case) form a tetrahedron. All H-C-H bond angles are exactly 109.5°, and all four

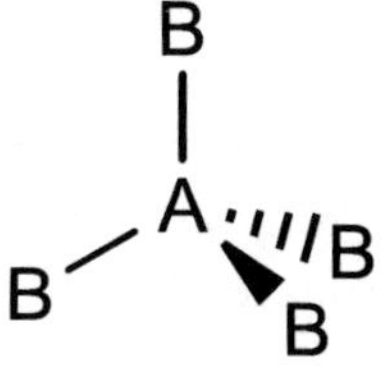

Tetrahedral

hydrogen's are in exactly the same positions relative to the carbon and the other three hydrogen atoms.

There are many examples of tetrahedral geometry: SO_4^{2-}, ClO_4^-, and SiF_4, to name a few.

2. One unshared electron pair: **AB₃U**

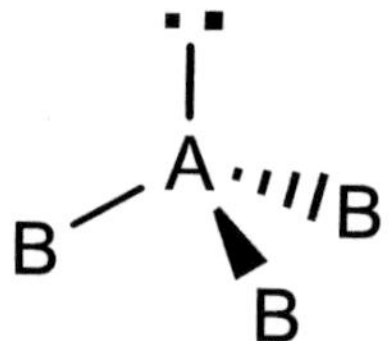

This is another common case. Consider NH_3. The four RHED around nitrogen (three single bonds and one unshared pair) form a tetrahedral electronic geometry, but the three shared

Trigonal Pyramid

RHED (with the hydrogen's on the end), will form a triangle, like in BF_3. But unlike BF_3, the central atom is __not__ in the same plane as the three hydrogens. This will look like a three-sided pyramid. This molecular geometry is called *trigonal pyramidal*. The unshared RHED will take up a bit more space, so the bond angle is somewhat less that 109.5°. (In NH_3, the bond angle is 107.3°)

3. Two unshared RHED: **AB₂U₂**

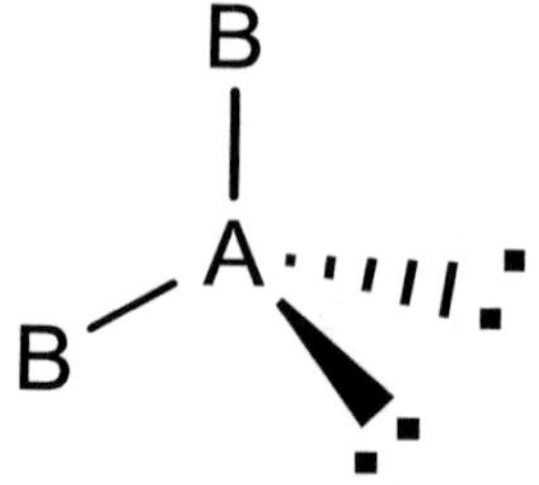

Another common case is H_2O. The four RHED in water are two single bonds and two unshared pairs on the oxygen. While the electronic geometry is still tetrahedral, the molecular geometry is *bent*. We've seen the "bent" structure above, but this is a little different. In the case of SO_2,

Bent

which was bent, the bond angle was just a little less than 120°, the trigonal planar bond angle. The unshared RHED in an AB_2U_2 case compress the H-O-H bond angle to something less than 109.5°. In the case of water, this angle is 104.5°.

D. Five Regions of High Electron Density

With five RHED there are at least 10 electrons around the central atom, so now these compounds exhibit extended valence shells, which can occur in some atoms in the third row and later rows. These are hypervalent compounds with an expanded octet. To envision what will happen with five RHED, go back to the tetrahedron, and think of four electron pairs around a

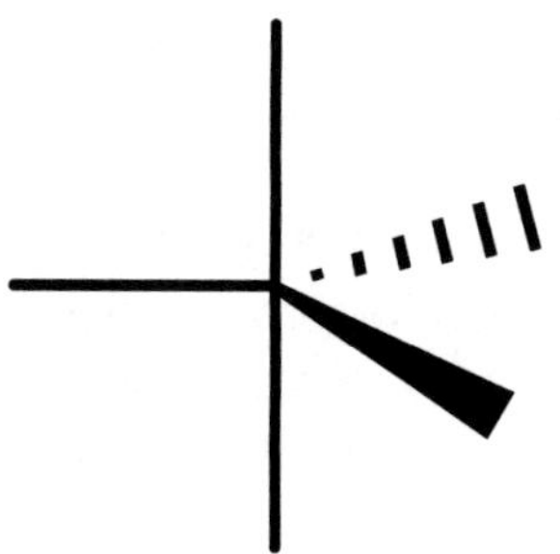

Trigonal Bipyramid

central atom. Now bring a fifth electron pair in perpendicular to the triangle formed by three of the original RHED. The fifth pair will force three of the originals back into the same plane. Now we have a set of three RHED forming a trigonal plane, and two more RHED forming a line perpendicular to the trigonal plane. The shape is like two trigonal pyramids placed base to base. This is called a *trigonal bipyramid*, the electronic geometry with five RHED.

Notice that we have two distinct kinds of RHED: the three in the trigonal plane, and the two perpendicular to the plane. The three in the trigonal plane are called "equatorial", and the two perpendiculars are called "*axial*". The equatorial positions have a different geometric relationship to the rest of the positions than do the axial positions.

1. No unshared pairs: **AB$_5$**

If there are no unshared RHED, the molecular geometry is the same as the electronic geometry, trigonal bipyramid. Probably the most famous example is PCl_5. Three of the Cl's are in the

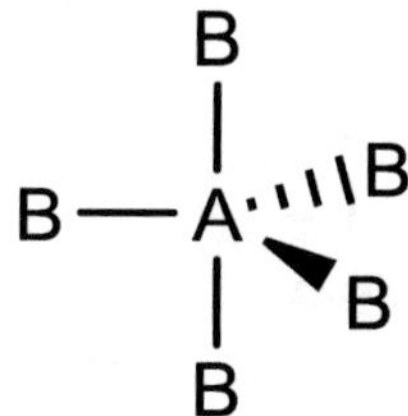

Trigonal Bipyramid

trigonal plane, forming 120° angles with each other. Two of the Cl's are axial, forming a bond angle of 180° with each other. The equatorial Cl form 90° angles with the axial Cl.

2. One unshared pair: **AB₄U**

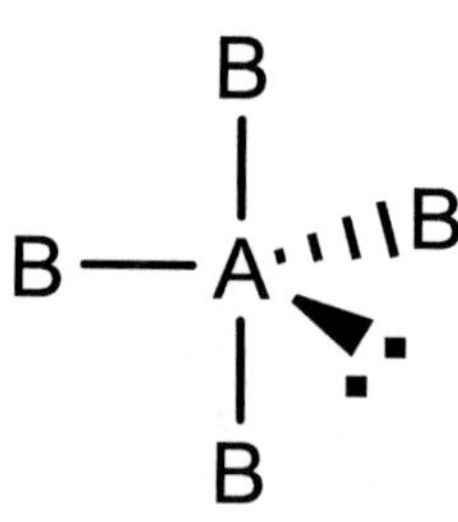

See-Saw

Now in determining the molecular geometry a new problem is encountered. There are two different positions for the unshared pair: equatorial and axial.

By Principle 2 we know that unshared pairs occupy more space than shared bonding pairs of electrons, but which position---axial, or equatorial---provides the optimum space? It turns out that there is _more space in an equatorial position than an axial position_. Therefore, the unshared pair will be in one of the equatorial positions.

An example of a molecule like this is SF_4. While the formula is something like CF_4, the lone pair on the sulfur gives it a different geometry. Two of the F are equatorial, the other two F are axial.

This unusual molecular geometry has been called **_disphenoidal_** or **_see-saw_**. The two equatorial F form an angle of a bit less than 120° with each other, and the two axial F form a bond angle of a bit less than 180°.

3. Two unshared pairs: **AB₃U₂**

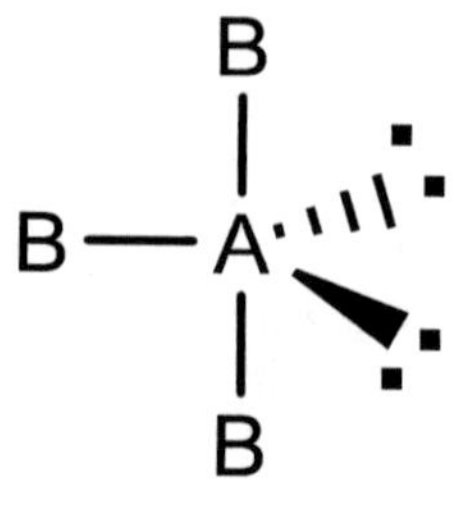

T-Shaped

Once again, we choose to put the unshared pair in an equatorial position. An example of this is ClF_3. One of the F is in the remaining equatorial position; the other two are axial.

The molecular geometry is **_T-shaped_**.

4. Three unshared pairs: **AB_2U_3**

The third unshared pair goes into the final equatorial position.
An example is KrF_2, one of the very few krypton compounds.
Both F are in the axial position, so the molecular geometry is strictly linear
Therefore, KrF_2 and $BeCl_2$ have the same molecular geometry, but for entirely different reasons.

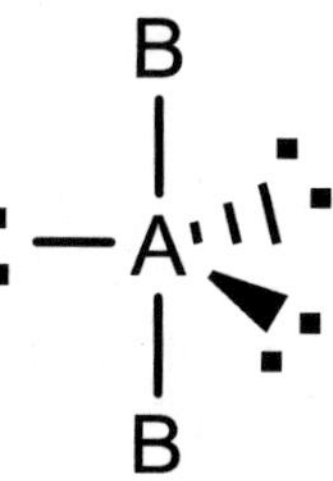

Linear

E. Six Regions of High Electron Density

Octahedral

The final case we'll consider here is six RHED around the central atom. Once again, in our imagination we'll start with the trigonal bipyramid. Imagine bringing yet another electron pair, this time bisecting two of the equatorial RHED. These two original equatorial RHED will be forced to move away, and in the end we'll have an *octahedral* electronic geometry.

The octahedral looks like two four-sided pyramids (square pyramids) back-to-back, and could be (but is not) called "square bipyramidal".

1. No lone pairs: **AB_6**

This molecular geometry is octahedral. An example is SF_6. All six F are entirely equivalent to each other, each forming a $90°$ angle with four other F, and an $180°$ angle with the fifth one.

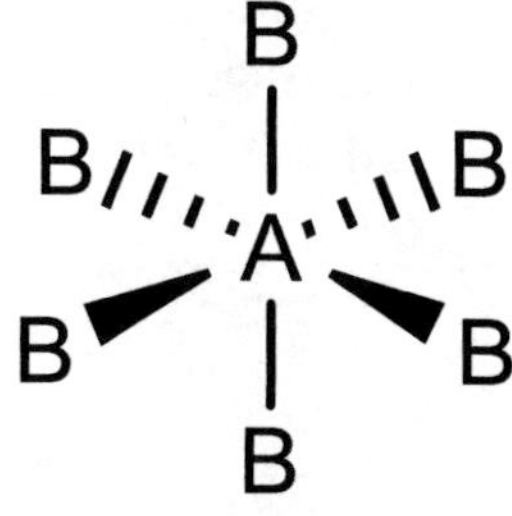

Octahedral

2. One lone pair: **AB$_5$U**

All positions in an octahedron are equivalent (unlike the trigonal bipyramid), and so we don't have to think about where the lone pair will go. It will simply assume one of the positions, and the remaining five atoms will lie in a molecular geometry, which is called *square pyramidal*.

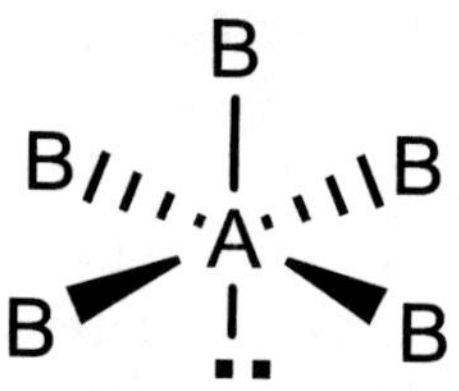

Square Pyramid

An example is the rare inorganic molecule IF$_5$.

3. Two lone pairs: **AB$_4$U$_2$**

Now there are two possible choices for the second lone pair: it can either be at 90° to the other lone pair, or at 180° to the other lone pair. Applying Principle 2, one assumes that the two lone pairs will be as far apart as they can be, and that will be at 180°.

Square Planar

An example is the noble gas compound XeF$_4$. The two lone pairs on Xe are on opposite sides of a plane of Xe and the four F atoms.

The four F atoms form a molecular geometry called *square planar*. The F atoms define the square, and the Xe lies directly in the center of the plane.

F. Summary

In the basic study of valence shell electron pair repulsion theory for common molecules using Lewis electron dot structures we therefore have five electronic geometries, and thirteen molecular geometries.

System	Total RHED	Unshared RHED	Shared RHED	Electronic Geometry	Molecular Geometry
AB_2	2	0	2	Linear	Linear
AB_3	3	0	3	Trigonal Planar	Trigonal Planar
AB_2U	3	1	2	Trigonal Planar	Bent
AB_4	4	0	4	Tetrahedral	Tetrahedral
AB_3U	4	1	3	Tetrahedral	Trigonal pyramidal
AB_2U_2	4	2	2	Tetrahedral	Bent
AB_5	5	0	5	Trigonal Bipyramidal	Trigonal Bipyramidal
AB_4U	5	1	4	Trigonal Bipyramidal	See-saw
AB_3U_2	5	2	3	Trigonal Bipyramidal	T-shaped
AB_2U_3	5	3	2	Trigonal Bipyramidal	Linear
AB_6	6	0	6	Octahedral	Octahedral
AB_5U	6	1	5	Octahedral	Square pyramidal
AB_4U_2	6	2	4	Octahedral	Square Planar

Electronegativity, Polarity, and Polar Molecules

In the 1930's and in his book *"The Nature of the Chemical Bond"* author and Nobel Prize winner *Linus Pauling* analyzed the energies of chemical bonds and he developed an *Electronegativity* scale. Electronegativity is defined as the ability of an atom to attract electrons to itself in a chemical bond. Atom electronegativities are useful in deciding if a chemical bond is polar, and which atom has a partial negative charge and which has a partial positive charge. The following scheme shows electronegativity values.

Pauling Electronegativity Values

2.2 **H**						
1.0 **Li**	1.6 **Be**	2.0 **B**	2.5 **C**	3.0 **N**	3.5 **O**	4.0 **F**
0.9 **Na**	1.3 **Mg**	1.6 **Al**	1.9 **Si**	2.2 **P**	2.6 **S**	3.1 **Cl**
0.8 **K**	1.0 **Ca**	1.8 **Ga**	2.0 **Ge**	2.2 **As**	2.6 **Se**	3.0 **Br**
0.8 **Rb**	0.9 **Sr**	1.8 **In**	1.9 **Sn**	2.0 **Sb**	2.1 **Te**	2.7 **I**
0.8 **Cs**	0.9 **Ba**					

One of the most confusing aspects of inorganic chemistry is the blurring of the line between ionic and covalent compounds. By looking at the periodic table and the associated electronegativity values of the elements we can predict if a compound has primarily covalent or ionic bonds.

If the electronegativity values of two atoms bonded together are exactly the same, we say it is a *purely covalent bond*. The only way that this occurs is to have two atoms of the same element bonded together, as in for example, Br_2, O_2, or N_2.

If the electronegativity value difference between two bonding atoms is near 1.0-2.0, then we say it is an ***ionic bond***. Examples would be LiF, NaCl, HF, etc. These are true salts or strong acids and bases.

If the electronegativity difference values are low, between 0 and 0.5, then we say the bond is a ***polar covalent bond***, but not purely covalent. One of the atoms has a larger electronegativity value than the other, and electron density is found to be greater on the atom with the highest electronegativity. This causes bonds to be polar.

Sometimes polar covalent bonds are arranged in a symmetrical fashion around the central atom, and we find that the molecule <u>does not</u> have a dipole moment. If the polar covalent bonds are unsymmetrically arranged around the central atom, we find that the molecule does have a ***dipole moment***.

Transition Metal Electronegativity Values

The transition metals have a very narrow range of electronegativity values that range from 1.3 for scandium to 1.9 for some of the heavier metals like Os, Ir, and Au. The electronegativity scale is therefore not as useful in describing the properties of transition metals as it is in describing the properties of ligands.

Pauling Electronegativity Values

3	4	5	6	7	8	9	10	11	12
21 Sc 1.3	22 Ti 1.4	23 V 1.5	24 Cr 1.6	25 Mn 1.6	26 Fe 1.7	27 Co 1.7	28 Ni 1.8	29 Cu 1.8	30 Zn 1.6
39 Y 1.3	40 Zr 1.3	41 Nb 1.5	42 Mo 1.6	43 Tc 1.7	44 Ru 1.8	45 Rh 1.8	46 Pd 1.8	47 Ag 1.6	48 Cd 1.6
71 La 1.3	72 Hf 1.3	73 Ta 1.4	74 W 1.5	75 Re 1.7	76 Os 1.9	77 Ir 1.9	78 Pt 1.8	79 Au 1.9	80 Hg 1.7

The low electronegativity values for the early transition metals means that they do not form as covalent an interaction with carbon in low oxidation states as do the heavier metals.

Dipole Moments

Certain molecules, in the gas phase, can be aligned in an electric field. Examples of these molecules are hydrogen fluoride and water. We should ask ourselves why this occurs…why are we able to align these molecules in an electric field?

HF Molecules Aligned in an Electric Field

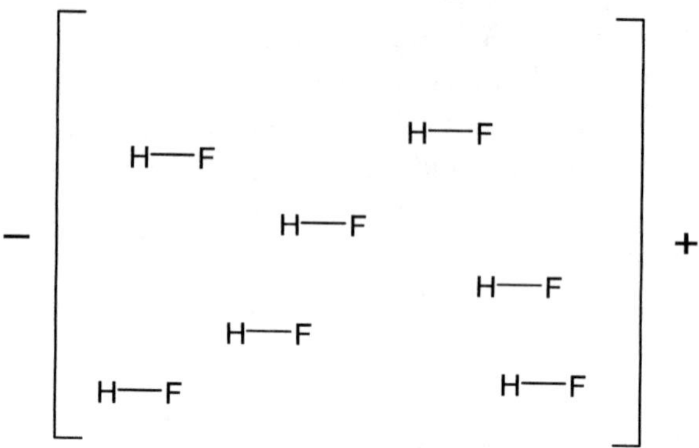

The hydrogen fluoride is a polar molecule, and this means that it has a dipole moment. The electronegativity of fluorine (4.0) is much greater than that of hydrogen (2.1), and thus the covalent bond between H and F is polar covalent since the difference in their electronegativity values is 1.9.

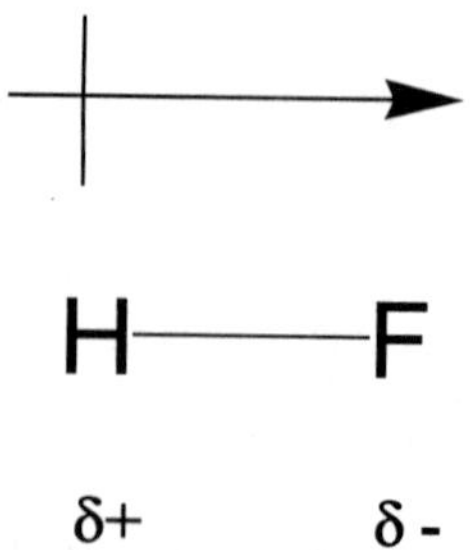

The arrow points to the negative end of the polar bond, and the + end of the arrow designates the positive end. The arrow indicates the polar nature of the bond, and also of the entire molecule since it is linear.

HF is not an ionic compound, therefore the negative end is considered to have a *"partial negative charge"*, which is designated with a ***delta minus*** (δ -) symbol, and the positive end likewise is considered to have a *"partial positive charge"* and is designated with a ***delta plus*** (δ+) symbol.

This is a case where the molecule contains a polar bond, and the molecule itself is polar and has a dipole moment. It is not always the case that a molecule contains polar bonds and is itself polar! Lets look at such a case in the next example.

To illustrate this point we will compare HF to another linear molecule, carbon dioxide or CO_2 . Carbon dioxide has double bonds between carbon and oxygen as can be seen in the structural representation shown next.

$$\delta - \qquad \delta+ \quad \delta+ \qquad \delta -$$

$$\longleftarrow + \quad + \longrightarrow$$

$$O = C = O$$

Each C=O bond is polar. However the polar bond of one C=O bond cancels out the effect of the other C=O bond with a net effect of <u>no polarity</u> for the entire molecule itself. (You can imagine these arrows as force vectors that cancel out)

The symmetrical nature of the molecule is the key to understanding if a molecule is polar or not. Symmetrical molecules, where any given one side of the molecule is the same as the other, are not polar, whereas unsymmetrical molecules are always polar.

The best way to systematically study polar molecules is to work up through the different molecular geometries. Since we have looked at linear molecules we should therefore examine trigonal planar molecules next, and then tetrahedral molecules.

The trigonal planar molecule boron trifluoride is not considered to be a polar molecule even though it contains three polar bonds.

This is illustrated in the representation of the molecule drawn below.

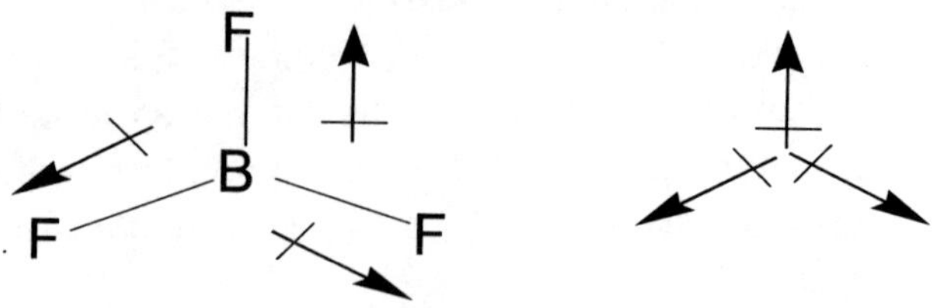

Again, this molecule is symmetrical and one-half of the molecule is like the other half.

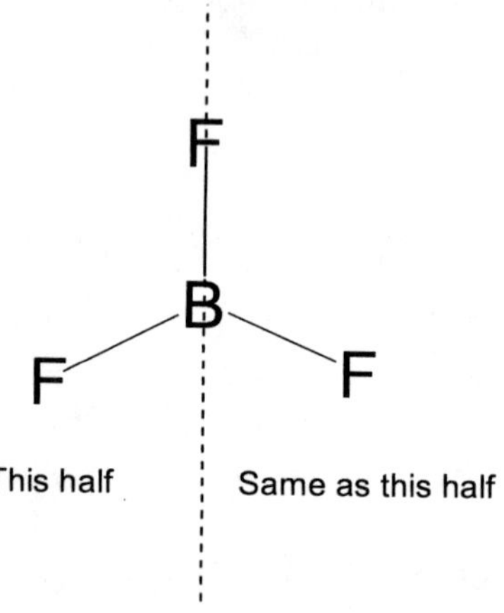

It is instructive to examine a planar molecule (BF$_2$Cl) that is very similar to BF$_3$.

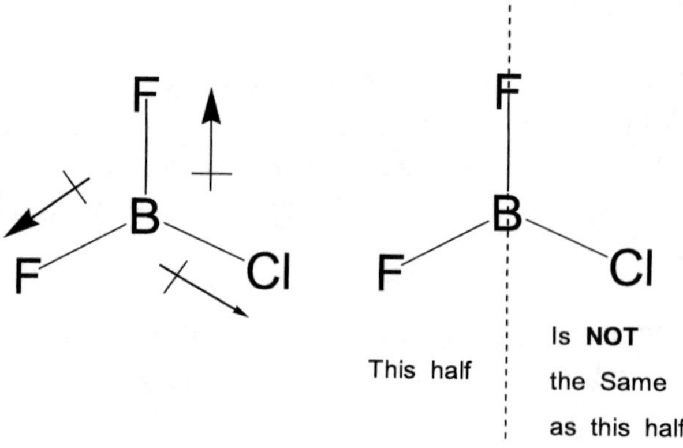

Notice that the arrow showing the polar bond between B and Cl is not as large in the drawing shown above as that of the polar bond between B and F.

This signifies that the B-Cl bond is not as polar as the B-F bond, and this can be calculated using the electronegativity values for each element.

The values are: B=2.0 , Cl=3.0 , and F= 4.0. Obviously the B-F bond is much more polar.

Next, the molecule is definitely unsymmetrical. As shown above, one-half of the molecule is not the same as the other half. With all these things considered, the BF_2Cl molecule shown above is a polar molecule and should exhibit a dipole moment, even if it is very small.

The scope of this book is limited, and is meant as an introduction to coordination chemistry on the sophomore college level. Thus, we use symmetry elements on a basic level, without introducing symmetry and group theory the way that it should really be presented (which would take an entire book to cover adequately) so the symmetry discussions in this section are meant to be basic and elementary.

Let us examine the next molecular geometry, the tetrahedral electronic geometry, which is more complicated.

Three molecules come easily to mind, methane (CH_4), methylene chloride (CH_2Cl_2) and carbon tetrachloride (CCl_4).

Methane and carbon tetrachloride are non-polar molecules because they have no dipole as can be seen in the previous scheme.

However, methylene chloride and chloroform do exhibit a dipole moment, and they are classified as ***polar organic molecules***.

The arrows in the following scheme depict polar covalent bonds, and also the overall dipole.

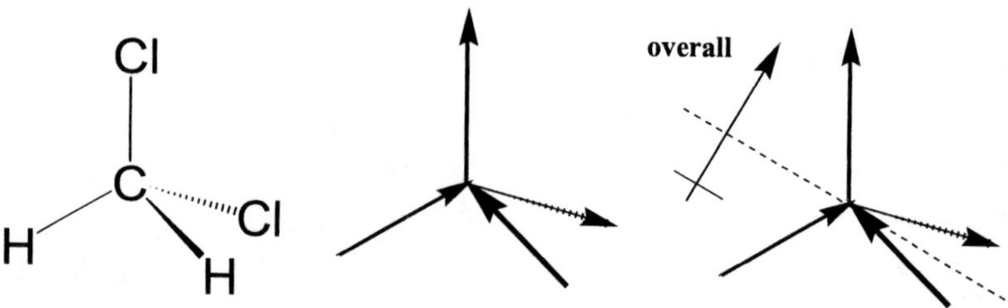

Dichloromethane has one side that is different from the other, creating an overall dipole moment.

The same is true for chloroform, but the dipole is slightly different.

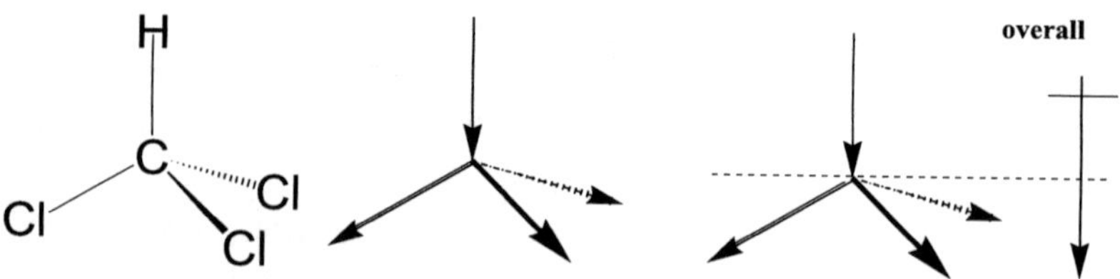

What about chloromethane $CClH_3$? Does that molecule have a dipole moment?

Yes, it does.

The trigonal bipyramid molecular geometry is actually two geometries in one. That is, there is a linear portion (axial), and also a trigonal planar portion (equatorial). Molecules that contain different substituents on the axial and equatorial positions are rare, but the mental exercise to see if a proposed molecule would have a dipole moment can be instructive.

The prototype molecule is PF_5, shown below.

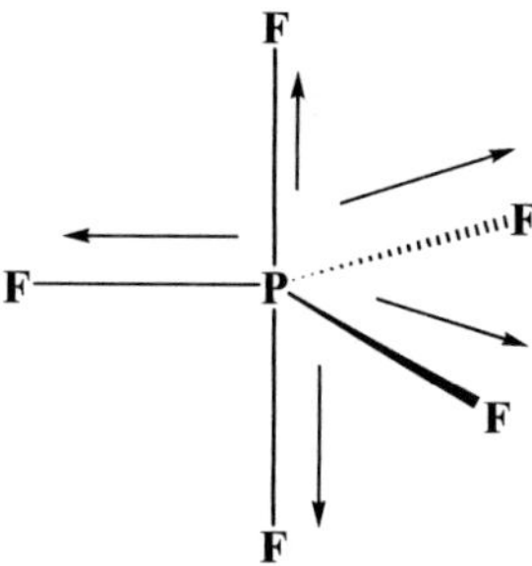

The molecule has no dipole moment as the vectors for all the polar bonds cancel out. If a Cl atom, in either an axial or an equatorial position, replaces one of the F atoms on PF_5 then the resulting molecule would be polar, with a weak dipole.

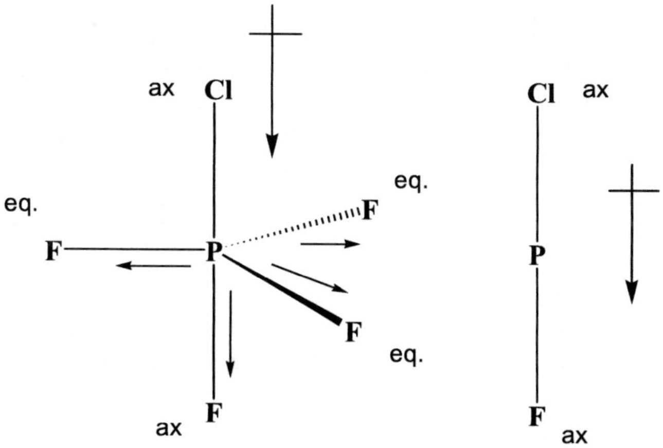

In the scheme above, the axial (ax) and the equatorial (eq) substituents on the molecule are labeled. Since the equatorial substituents are all F, with polar P-F bonds, they cancel each other out.

On the other hand, the axial substituents are different, making one side of the molecule different from the other. The P-F bond is more polar than the P-Cl bond, so the dipole lies in the direction indicated.

If the Cl were placed on an equatorial position instead, the following situation would occur as depicted in the next scheme.

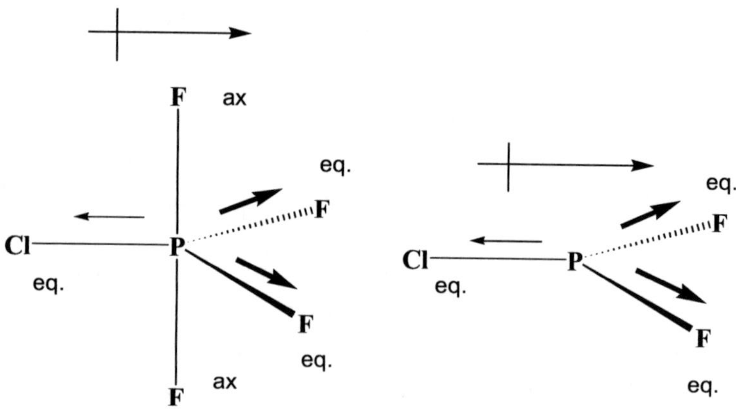

In the scheme above, since the axial substituents are the same, all F, with polar P-F bonds, then they cancel each other out. On the other hand, the equatorial substituents are different, making one side of the molecule different from the other. The P-F bond is more polar than the P-Cl bond, so the dipole lies in the direction indicated.

The disphenoidal geometry is, by its very unsymmetrical nature, prone to exhibit a dipole moment in all compounds that adopt that structure. For example the molecule SF_4 has a dipole moment, because we can draw a plane through the molecule where one-half is different from the other.

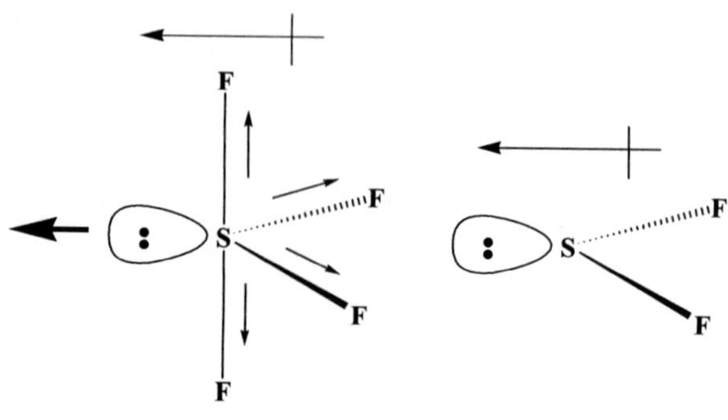

Disphenoidal

Molecules with a T-Shaped molecular geometry are, of course, polar molecules. ClF_3 is such an example, and the following scheme depicts the molecule. As in the previous example, lone pairs of electrons are considered to be regions of high electron density (RHED), so they are considered strongly when proposing dipole moments.

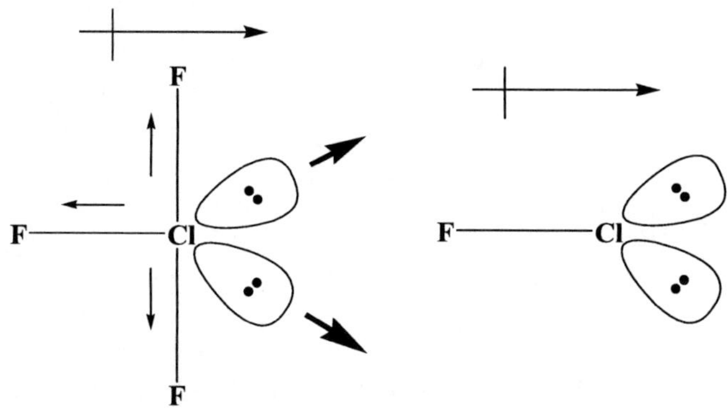

T-Shaped

The octahedral geometry is an interesting case since it could potentially involve many of the transition metal compounds that we may observe. Again, it is the breaking of perfect symmetry about the molecule that is the key to observation of a dipole moment. The molecule SF_6 is the prototype molecule of octahedral molecular geometry for a non-transition metal compound, and it is not a polar molecule, as can be seen in the scheme below.

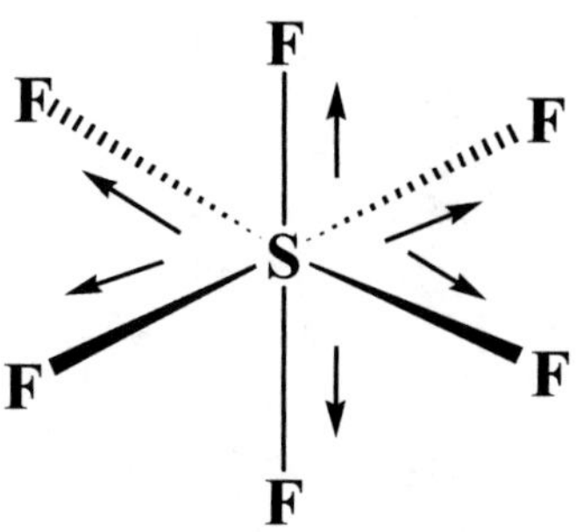

If one examines a metal complex that does not have this perfect symmetry around the central atom, for example $[Co(NH_3)_5Cl]^{2+}$, then the molecule does exhibit a dipole moment.

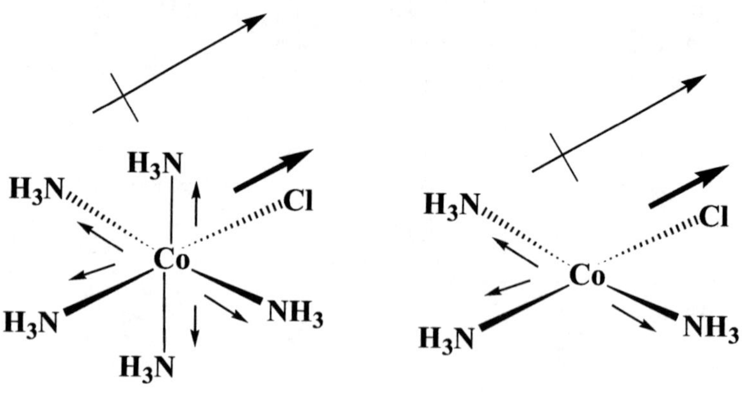

This is also the case with cis-$[Co(NH_3)_4Cl_2]^+$, where it does have a dipole moment, but this is not the case for trans-$[Co(NH_3)_4Cl_2]^+$, which does not have a dipole moment.

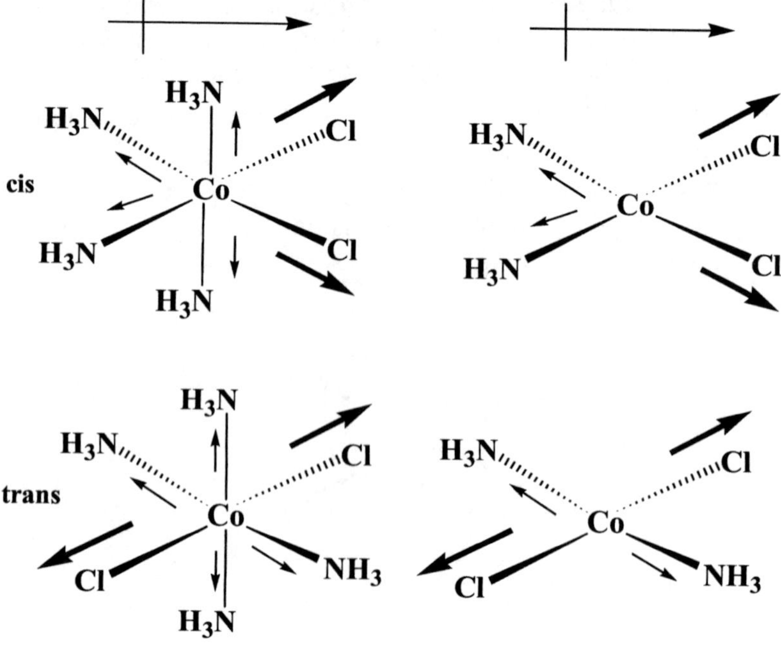

If a compound is examined with an octahedral electronic geometry such as IF_5, which is a rare example of a non-transition metal compound with a square pyramid molecular geometry, one sees that it has a dipole moment.

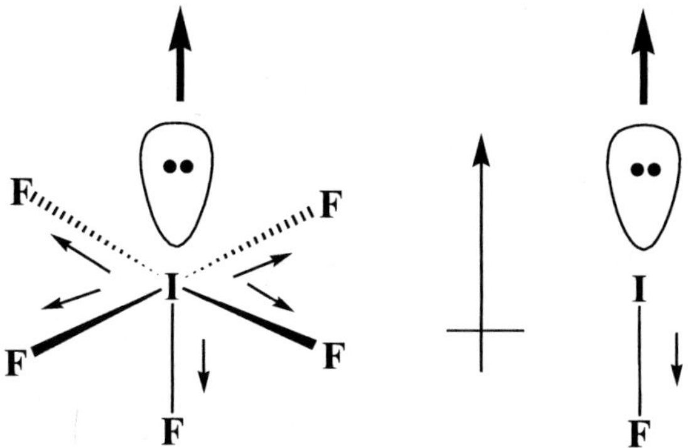

The square planar molecular geometry is basically a special case of the octahedral molecular geometry. In the square planar case of non-transition metal compounds such as the rare XeF_4 molecule, there is no dipole moment.

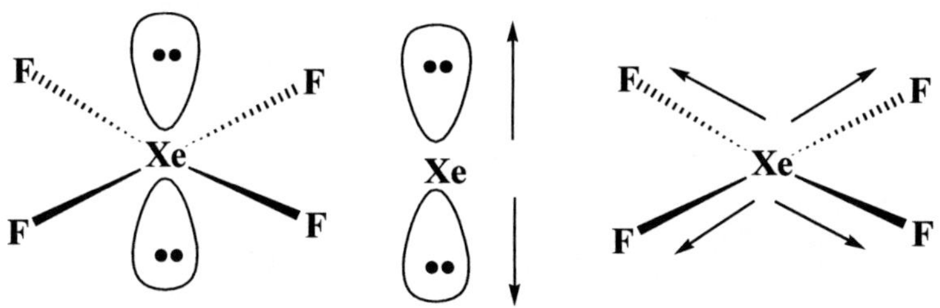

The electron pairs cancel each other out (axial positions) and the Xe-F bonds cancel each other out (equatorial positions) leaving a molecule with no discernible dipole moment.

In the case of transition metal complexes that do not have lone pairs associated with the central atom in a VSEPR sense, the situation is different, but there are similarities. For example, $[PtF_4]^{2-}$ has no dipole moment, but $[Pt(NH_3)_3Cl]^-$ does.

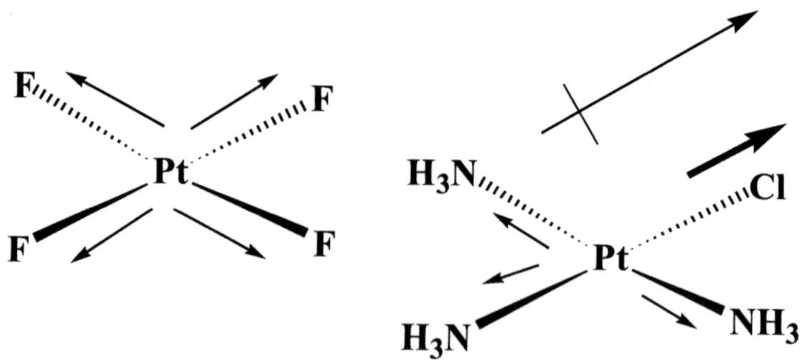

The molecule cis-platin has a dipole moment. (Compare to cis-$[Co(NH_3)_4Cl_2]^+$) The isomer of cis-platin, the trans-platin molecule, has no dipole moment. The cis-and trans-platin molecules are shown in the scheme below.

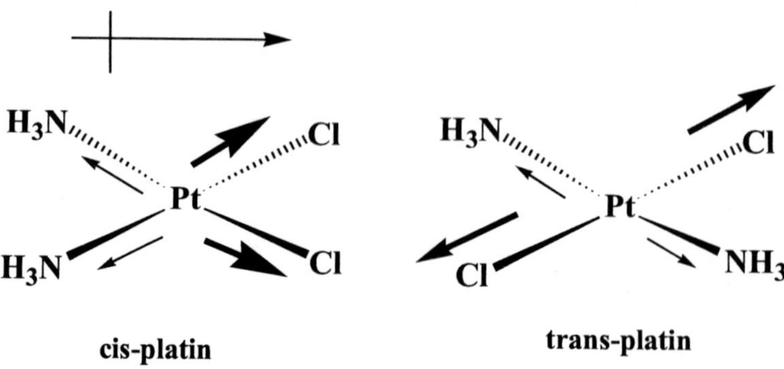

cis-platin

trans-platin

Rare Geometries

There are higher coordinate geometries than octahedral (six coordinate) but they are rare, with few examples. Even coordination number six has a rare representative, the trigonal prismatic geometry.

The best example of this rare geometry is $Re[S_2C_2(C_6H_5)_2]_3$, a rhenium dithiolate complex, where the dithiolate ligand is a ***bidentate ligand*** (see Chapter 3), which forces the Re into this geometry.

Seven, eight, and nine coordinate metal complexes are known. Usually, to get this many ligands around a central atom, the ligands need to be small (like F^-) and the metal has to be fairly large (like a 2nd or 3rd row transition metal, or a rare earth or actinide metal).

Coord #	Geometry	Example
7	Pentagonal bipyramid	ZrF_7^{2-}, $V(CN)_7^{4-}$
7	Monocapped octahedron	NbF_7^{2-}, TaF_7^{2-},
7	Monocapped trigonal prism	$NbOF_6^{3-}$
8	Cubic	$Na_3[PaF_8]$, $Na_3[UF_8]$
8	Square antiprism	ReF_8^{2-}, $W(CN)_8^{4-}$ $Zr(acac)_4$
8	Dodecahedron	$Mo(CN)_8^{4-}$
8	Hexagonal bipyramid	$[CdBr_2(18\text{-crown-}6)]$
9	Tricapped trigonal prism	ReH_9^{2-}

High Coordinate Geometries

7-coordinate
Pentagonal Bipyramid

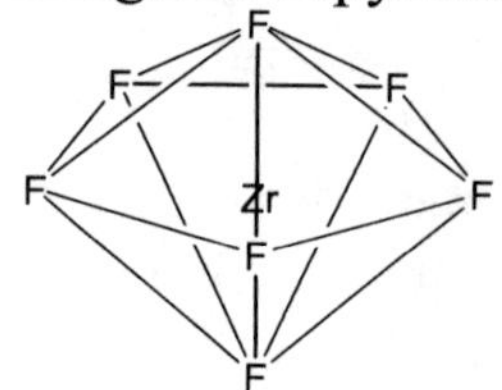

7-coordinate
Monocapped Octahedron

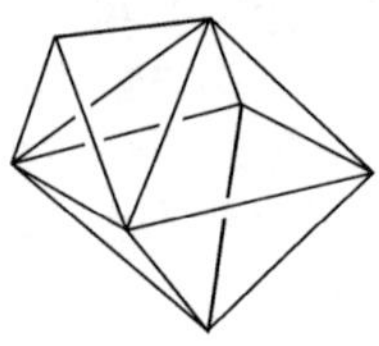

7-coordinate

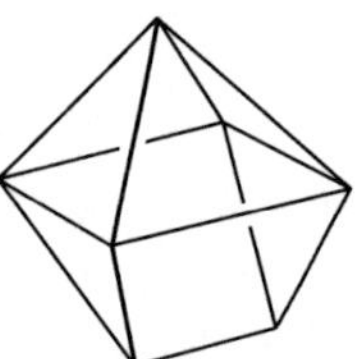

Monocapped Trigonal Prism

8-coordinate
Cubic

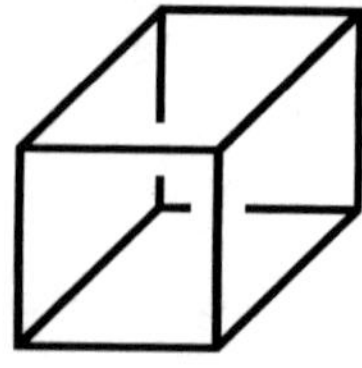

8-coordinate
Square Antiprism

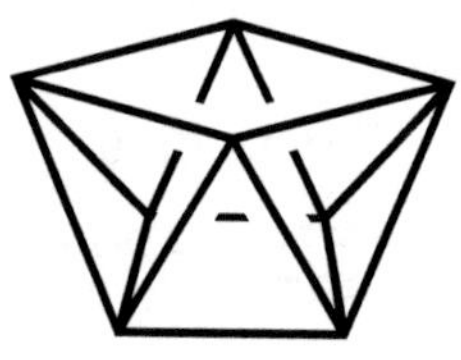

8-coordinate

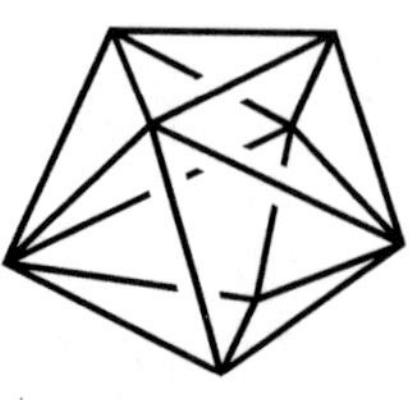

Dodecahedron

8-coordinate

Hexagonal Bipyramid

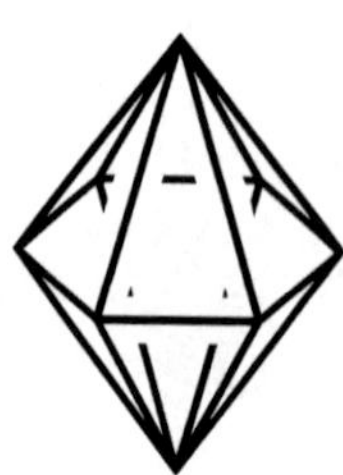

9-coordinate

Tricapped Trigonal Prism

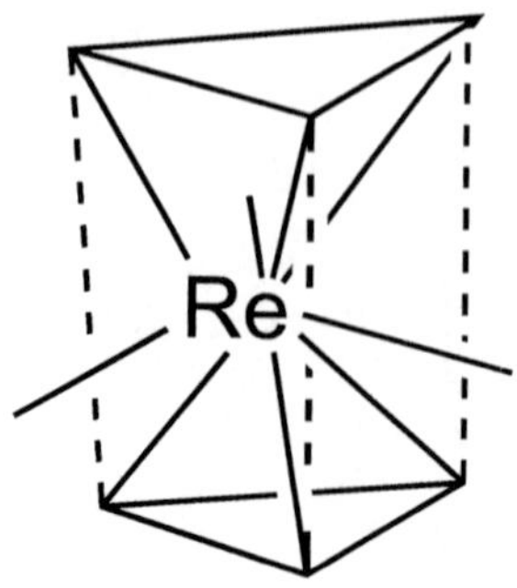

Terms and Definitions Chapter 2

Lewis Electron Dot Structures –

G.N. Lewis --

VSEPR Theory –

The Octet Rule –

Bonding Pairs Of Electrons –

Lone Pairs Of Electrons –

Multiple Bond –

Resonance Structures –

Delocalized –

Formal Charge –

Hypervalent Compounds –

Electron Deficient –

Lewis Acid –

Lewis Base –

Central Atoms –

Peripheral Atoms –

Regions Of High Electron Density (RHED) –

Electronic Geometry –

Molecular Geometry –

Linear –

Trigonal Planar –

Tetrahedral –

Trigonal Pyramidal –

Bent –

Trigonal Bipyramid –

Disphenoidal (Seesaw) –

T-Shaped –

Axial And Equatorial Sites –

Octahedral –

Square Pyramid --
Square Planar –

Linus Pauling --
The Nature Of The Chemical Bond --
Electronegativity –
Valence Bond Theory ---
Purely Covalent Bond –

Ionic Bond --
Polar Covalent Bond --

Dipole Moment –

Partial Negative Charge --
Partial Positive Charge --

Polar Organic Molecules

Chapter 2 Problems and Exercises

1. Which of the following species can have resonance structures?

    (A) O_3 (B) BF_4^- (C) CO_3^{2-}
    (D) NO_2^- (E) CH_3NH_2

2. How many resonance forms can be written for the nitrate ion?

3. How many lone pairs of electrons does the nitrogen atom possess in the Lewis structure of HCN?

4. For the Lewis structure below, the formal charges on N, C, and S, respectively, are...

$$[::C=S=N::]^-$$

5. For the Lewis structure below, the formal charges on C, S, and N, respectively, are....

$$[::N=C=S::]^-$$

6. All of the following have Lewis structures that obey the octet rule except ...

(A) NO   (B) NO_2^+ (C) N_2O_4 (D) N_2O (E) N_3^-

7. The number of lone pairs around the central iodine atom in I_3^- is _______.

8. Which of the following are Lewis acids?

    CN^-, Al^{3+}, Cl^-, PCl_3, BF_3

9. What is the molecular shape of $AlCl_4^-$?
10. What is the molecular shape of IF_2^-?

11. What is the molecular shape of SO_3?

12. The F–S–F bond angle in SF_2 is …

13. What is the molecular shape of $CO_3{}^{2-}$?

14. The Cl–P–Cl bond angle(s) in PCl_5 is/are ….

15. Which one of the following molecules is polar?
Draw them all.
  (A) COS (B) S_2 (C) CO_2 (D) CS_2 (E) O_2

16. Which one of the following molecules is polar?
Draw them all.
  (A) CH_2F_2 (B) XeF_4 (C) CH_4 (D) BF_3 (E) CCl_4

17. All of the following molecules are polar except…
Draw them all.
  (A) SF_2 (B) XeF_2 (C) OF_2 (D) SF_4 (E) NOCl

18. Which of the following ions have noble gas electron
configurations? List the valence shell electronic
configurations for all of them.
  (a) Fe^{2+}, Fe^{3+}, Sc^{3+}, Co^{3+}
  (b) Tl^{+}, Te^{2-}, Cr^{3+}
  (c) Pu^{4+}, Ce^{4+}, Tl^{3+}
  (d) Ba^{2+}, Pt^{2+}, Mn^{2+}

19. Write Lewis electron dot structures that obey the octet
rule for each of the following:
(a) CN-   (b) CO   (c) PH_3 (d) PF_3 (e) SCN-

20. Although both $Br_3{}^-$ and $I_3{}^-$ ions are known, the $F_3{}^-$ ion
has not been observed.
        Why not?

System	Total RHED	Unshared RHED	Shared RHED	Electronic Geometry	Molecular Geometry
CS_2					
BF_3					
CO_3^{2-}					
O_3					
SO_4^{2-}					
PCl_3					
ClO_3^-					
H_2S					
PF_5					
PF_4^-					
ICl_3					
XeF_2					
PF_6^-					
BrF_5					
XeF_4					

Practice Quiz   *DRAW STRUCTURES!!!* That will help you!

_____1. Which of these molecules do contain polar bonds?
(a) H_2O (b) H_2S (c) CO (d) CO_2 (e) all contain polar bonds

_____2. Which of these molecules do not contain polar bonds?
(a) H_2 (b) N_2 (c) Ar (d) I_2 (e) none contain polar bonds

_____3. Which of these molecules DOES NOT have a dipole moment?
 (a) H_2 (b) BF_3 (c) Ar (d) I_2 (e) none have a dipole moment

_____4. Which of these molecules DOES have a dipole moment?
(a) CH_4 (b) XeF_4 (c) O_2 (d) CO_2 (e) NH_3

5.-9. Match the compound with its correct ***electronic geometry*** (one each)
(a) CO_2 (b) H_2O (c) PCl_5 (d) XeF_4 (e) NO_2^-
_____5. Tetrahedral _____6. Trigonal Bipyramidal
_____7. Octahedral
_____8. Trigonal Planar
_____9. Linear

_____10. Triple bonds count as only a single pair of electrons when deciding the correct electronic geometry around the central atom of a molecule.   (a) true (b) false

11.-15. Match the compound with its correct ***molecular geometry*** (one each)
(a) CH_4 (b) H_2O (c) PCl_5 (d) XeF_4 (e) SF_4

_____11. Tetrahedral   _____12. Trigonal Bipyramidal
_____13. Square Planar   _____14. Bent
_____15. See-Saw

_____16. The electronic geometry of the central atom in PCl_3 is:
(a) pyramidal (b) trigonal planar (c) tetrahedral
(d) octahedral (e) trigonal bipyramidal

_____17. The **molecular geometry** of PCl_3 is:
(a) pyramidal (b) trigonal planar (c) tetrahedral
(d) octahedral (e) trigonal bipyramidal

_____18. Four of the following statements about the ammonia molecule, NH_3, are correct.
One is not correct. Which is it?
(a) The ammonia molecule has tetrahedral molecular geometry.   (b) Since nitrogen is more electronegative than hydrogen, the bond dipoles are directed toward the nitrogen atoms.
(c) The bond dipoles re-enforce the effect of the unshared pair of electrons on the nitrogen atom. (d) The bond angles in the ammonia molecule are less than $109°$.

_____19. The **molecular shape** of SO_3 is:
   (a) pyramidal (b) trigonal planar (c) tetrahedral
   (d) octahedral (e) trigonal bipyramidal

_____20. **The molecular shape** of $AlCl_4-$ is:
   (a) pyramidal (b) trigonal planar (c) tetrahedral
   (d) octahedral (e) trigonal bipyramidal

_____21. The Lewis dot formula for Br_2 shows _____.
(a) a single covalent bond.   (b) a double covalent bond.
(c) a triple covalent bond.   (d) a single ionic bond.
(e) a total of 8 x 2 = 16 electrons.

_____22. The number of unshared pairs of electrons in the outer shell of sulfur in H_2S is _____.
   (a) one   (b) two (c) three (d) four   (e) zero

____23. The number of unshared pairs of electrons in the outer shell of arsenic in AsF_3 is __.
  (a) one   (b) two   (c) three   (d) four   (e) zero

____24. Which molecule exhibits resonance?
(a) BeI_2 (b) O_3 (c) H_2S (d) PF_6 (e) CO_2

____25. How many lone pairs of electrons are there on the Xe atom in the XeF_4 molecule?
(a) one (b) two (c) three (d) four (e) zero

____26. The elements of Group VIIA may react with each other to form covalent compounds. Which of the following single covalent bonds in such compounds is the most polar bond?
(a) F-F   (b) F-Cl   (c) F-Br   (d) F-I   (e) Cl-I

____27. Which one of the compounds below has the bonds that are the most polar?
(a) H_2S (b) PH_3 (c) $AsCl_3$ (d) SiH_4 (e) $SbCl_3$

____28. Which one of the following molecules does not have a dipole moment?
(a) BrCl   (b) FCl   (c) FBr   (d) O_2 (e) ClBr

Chapter 3. Ligands and Nomenclature

The transition metal studies of Werner were followed later by the concepts of G. N. Lewis and N. V. Sidgwick, who recognized that formation of a chemical bond required the sharing of an electron pair. This led to the idea that a molecule or ion with an electron pair can donate these electrons (definition of a *Lewis base*) to a metal ion, which is considered to be an electron acceptor (definition of a *Lewis acid*).

$$H^+ \quad + \quad \left[:\overset{\cdot\cdot}{\underset{\cdot\cdot}{O}}-H \right]^- \quad \longrightarrow \quad H_2O$$

Lewis Acid   Lewis Base

$$BF_3 \quad + \quad :NH_3 \quad \longrightarrow \quad BF_3-NH_3$$

Lewis Acid   Lewis Base   Coordinate Covalent Bond

The definition of a *ligand* is any molecule or ion that has <u>at least one electron pair that can be donated to a metal</u>.

Ligands are *Lewis bases* since they donate a pair of electrons to the metal and are attracted to them, thus they are also called *nucleophiles*. Molecules such as BF_3 with incomplete valence electron shells, or metal ions in a metal complex, are *Lewis acids* since they are electron acceptors, and they are also *electrophiles* since they seek out ligands.

When one atom donates both electrons to form a covalent bond, it is termed a *coordinate covalent bond*. Coordinate covalent bonds are formed between ligands and metals when metal complexes are formed.

$$Ni^{2+} \quad + \quad 4 \left[:\overset{\cdot\cdot}{\underset{\cdot\cdot}{Cl}}: \right]^- \quad \longrightarrow \quad \left[NiCl_4 \right]^{2-}$$

Lewis Acid   Lewis Base

$$Ni^{2+} \quad + \quad 4 \; :NH_3 \quad \longrightarrow \quad \left[Ni(NH_3)_4 \right]^{2+}$$

Lewis Acid   Lewis Base

Hard and Soft Acid-Base Concepts

The use of hard and soft acid-base concepts is more of a rule of thumb that is useful in describing the general tendencies of certain metals *(Lewis Acids)* to bond preferentially to certain ligands (*Lewis Bases*) than it is a real working theory.

In general there is a tendency for ligands to bind to metals to form the most stable complex, so that a ligand that binds weakly to a metal may be replaced by a ligand that binds more strongly. (L* is a more strongly binding ligand than L in this example)

$$\text{M-L} + \text{L*} \leftrightarrow \text{M-L*} + \text{L}$$

A ranking of the bonding of ligands with a certain metal species can be determined in order to construct a ranking of ligand Lewis base strengths. This has been done, of course, and it turns out that the ranking of ligand Lewis base strengths depends on the metal.

In general, ligands such as NH_3, OH^-, Cl^-, F^-, O^{2-}, and H_2O tend to form strong complexes with metals such as Cr^{3+}, Fe^{3+}, Co^{3+}, and Ti^{4+}. These ligands all have ligating atoms (O, N, etc.) that are all very electronegative and small. The metals are also very small, thus the ligands and metals are both considered ***nonpolarizable*** and ***hard***.

On the other hand, ligands such as CN-, PR_3, AsR_3, SR_2, SR-, and CO tend to form strong complexes with metals such as Cu^+, Hg^+, Cd^{2+}, and Au^+. These ligands all have ligating atoms (S, P, etc.) that are all of a lower electronegativity than the hard ligands discussed above, and they are larger. The metals are also much larger, thus the ligands and metals are both considered ***polarizable*** and ***soft***.

The terms soft and hard are very imprecise but the concept is useful for guessing the stability of metal complexes with various ligands. The ambiguity is often due to the large number of borderline cases of ligands and metals.

Ligands can be classified in several ways.
(A) Ligands may be classified by <u>bonding mode</u>
(B) Ligands may be classified by <u># of donated electrons</u>
(C)Ligands may be classified as <u>neutral, anions or cations</u>
(D) Ligands may be classified by <u>structural type</u>

(A) There are *two main classes of <u>ligand bonding types</u>*
according to Cotton and Wilkinson (in "Advanced Inorganic
Chemistry"), and for the general introductory study of
transition metal coordination compounds this is surely the
most important way to classify ligands. The reason that
this is the most important way to classify ligands is not
apparently obvious to the student of coordination chemistry
at this point, but later, after the discussion of crystal field
theory and molecular orbital theory, the importance will
become apparent. <u>The two bonding modes are discussed
below, along with representative ligands</u>.

(1) Classical or simple donor ligands act as electron-pair
donors to acceptor ions or molecules, and form complexes
with all types of metal ions. These classical ligands are
common neutral or anionic ligands such as chloride ions,
water molecules, and ammonia molecules, but they may be
more complicated ligands such as EDTA. These ligands are
simple sigma electron pair donors.

(2) Nonclassical ligands, π-bonding or π-acid -ligands,
form coordination compounds only with transition metal
atoms. There are three major ligand groups of this type:
**(a) triply bonded molecules: carbon monoxide (CO),
cyanide (CN)$^-$, nitrosyl ion (NO$^+$), isocyanides (CN-R)**
(b) unsaturated organic molecules like alkenes and
alkynes, benzene or the cyclopentadienyl anion,
(c) organophosphine ligands such as triphenylphosphine
and bis(diphenylphosphino)ethane (includes arsines).
These ligands utilize some form of π-bonding to
enhance their sigma donor bonding to a metal ion, thereby
increasing the overall strength of the metal-ligand bond.

Unfortunately, as will be discussed later in the text, the borderline between sigma and pi bonding between ligands and metals is not as clear as one would like it to be.

(B) Ligands may also be classified according to the number of electrons that they contribute to a central atom when these ligands _are regarded as neutral species._ This is a matter of _convention_ on how the electrons are counted (p121-124). Are the electrons counted as on the ligand or the metal or both? The differences cause confusion.

Anionic ligands that can form a covalent bond are considered to be **one-electron donors** - examples are Cl^-, OH^-, and CH_3^-. These one-electron donors have a negative charge since they have, in effect, oxidized the metal, or could be considered to be sharing one electron with the metal.

To be counted as a one-electron donor means that one considers that the _ligand is bonding covalently_ with the metal, _not in an ionic fashion._ Later, we will examine both of these methods (ICC method and the NCC method) in counting electrons in Chapter 5. The ICC method is preferred by most chemists and **does not** distinguish anionic ligands as one-electron donors but as two-electron donors.

Any compound with an electron pair is a **two-electron donor.** Common examples are neutral molecules such as ammonia ($:NH_3$) or water ($H_2O:$).

Groups that can form a single bond and at the same time donate a pair of electrons can be considered to be **three-electron donors.** For example, the acetate ion can be either a one- or a three-electron donor.

A molecule with two electron pairs (for example, ethylene diamine, $H_2NCH_2CH_2NH_2$) can be regarded as **a four-electron donor,** and so on.

(C) A third way of classifying ligands is by their electrical charge. Ligands are <u>neutral, anions or cations.</u>

By convention, neutral ligands do not interfere with the oxidation state of the metal. On the other hand, anionic ligands oxidize the metal by one for every negative charge they possess, and cationic ligands decrease the oxidation state of the metal by one for every positive charge they possess. <u>This is the ICC basis for electron counting. (p121)</u>

(D) A fourth way of classifying ligands is _structurally,_ which is by the number of individual bonding connections that the ligand makes to the central metal atom.

When only one atom becomes bonded the ligand is said to be **_monodentate_ or _unidentate,_** such as the ligands in $Co(NH_3)_6{}^{3+}$, $NiCl_4{}^{2-}$, or $Ru(CN)_6{}^{3-}$. Thus, even though the ammonia (NH_3) ligand has four atoms, there is only one atom which bonds to the metal; the ligating atom is therefore nitrogen (N).

When a ligand uses two atoms to bond to a metal center, it is **_bidentate_**. This is the case for the tertiary organophosphine ligands, dppe= bis(diphenylphosphino)methane, and dppe= bis(diphenylphosphino)ethane, shown below:

The organophosphine ligands, as depicted in the last pages of Table 3.1, are a very important class of ligands for transition metal complexes, the structures of which can be tailored to suit the needs of the chemist.

Bidentate ligands bonded to only one metal atom are termed **chelate**, *(key-late--from the Greek--like a crabs claw)* as in the example using the **bidentate chelating** ethylene diamine ligand, or as shown below with the dppe ligand.

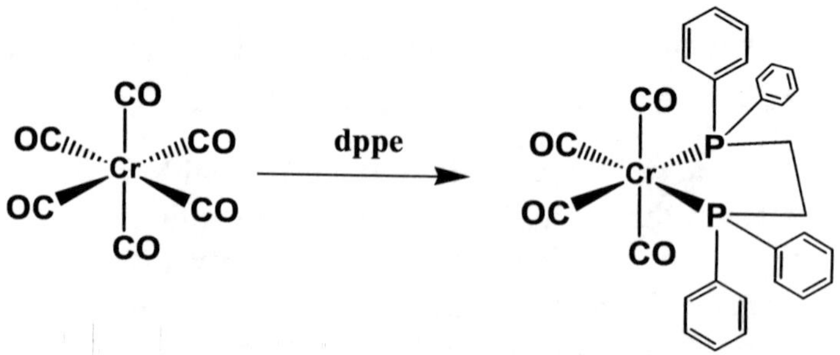

A ligand may also be **tridentate, tetradentate,** *or* **hexadentate,** and so on, these are all **multidentate ligands**.

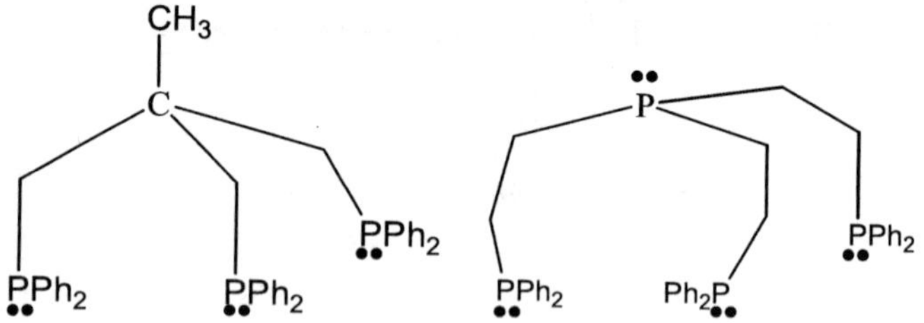

The multidentate phosphine ligands shown above can act as tridentate and tetradentate ligands respectively.

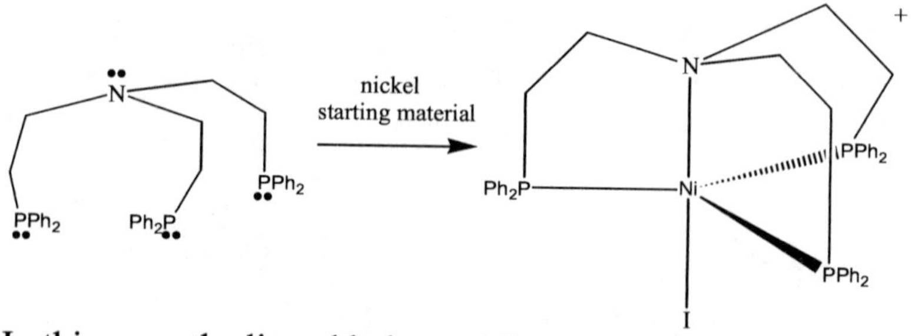

In this case, the ligand helps stabilize an odd geometry for the nickel metal complex, the trigonal bipyramid.

The following schemes show how multidentate ligands bind with a transition metal through their ligating atoms (in bold). en is a bidentate ligand

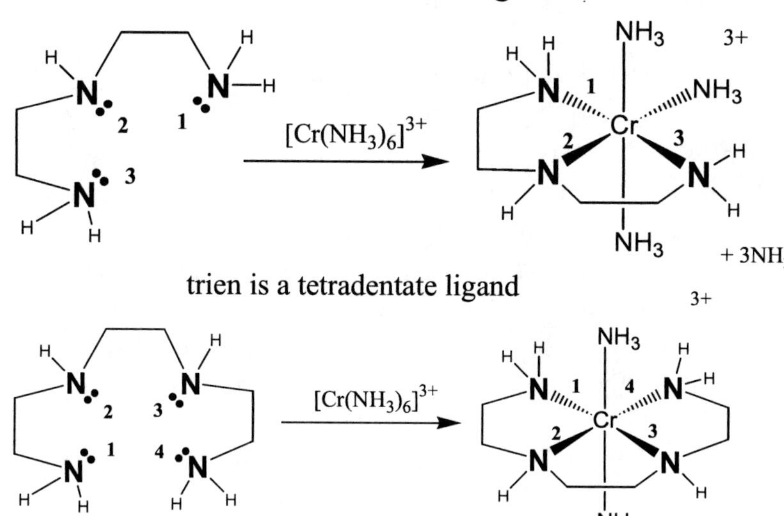

dien is a tridentate ligand

trien is a tetradentate ligand

EDTA can be a hexadentate ligand

The Chelate Effect

Chelating ligands often form very stable compounds with metal atoms because of ***inherent thermodynamic stability***. This thermodynamic stability is called the ***chelate effect***, and has enthalpic and entropic components.

The entropic component is by far the most important portion of the chelate effect and can be seen in the following scheme:

$$[Ni(NH_3)_6]^{2+} + 3\ en \longrightarrow [Ni(en)_3]^{2+} + 6\ NH_3$$

4 molecules   **produce**   **7 molecules**

The reaction produces more molecules than were originally reacted together, thus increasing the entropy (ΔS) and contributing to a negative free energy (ΔG). Remember the equation learned in general chemistry: $\Delta G = \Delta H - T\Delta S$, where ΔG is Gibbs free energy, ΔH is enthalpy, and ΔS is entropy. If ΔG is negative, then the process is spontaneous.

If one "arm" of the ligand is knocked away from the metal atom, in a collision in solution for example, then the other arm holds the chelate together long enough for the other arm to reattach itself to the metal. This is an enthalpic stability in that the total energy of the chelate metal-ligand bond is double that of a comparable monodentate ligand.

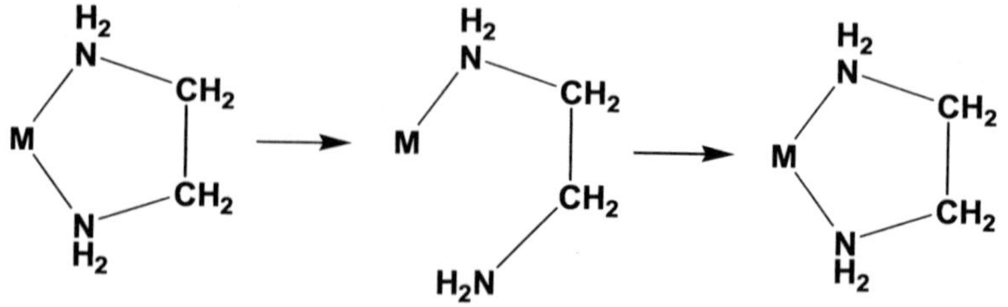

Many ligands act as ***bridging groups***. In many cases single atoms or anions act as ***unidentate bridging*** ligands. This means that there is only *one* ligand atom that forms two or three bonds to different metal atoms. Larger, multi-atom ligands can sometimes act as ***bidentate bridging*** ligands, as seen in the following examples.

Unidentate Bridging   **Bidentate Bridging**

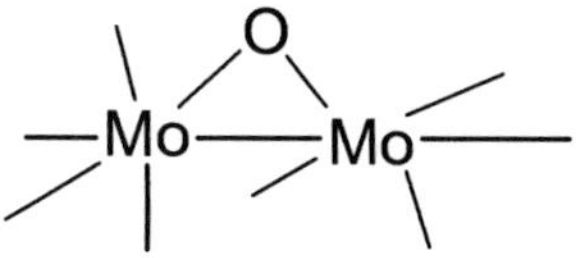

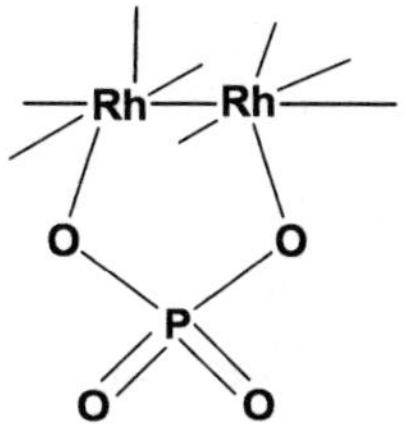

For monoatomic ligands, such as the chloride ion, and ligands containing only one donor atom, the ***unidentate*** form of bridging is the only possible mode.

Bridging ligands can be used to construct complexes and organometallic catalysts that exhibit interesting properties. In the following example, the rhodium complex has an open space between the Rh atoms that provides for interesting catalytic activity.

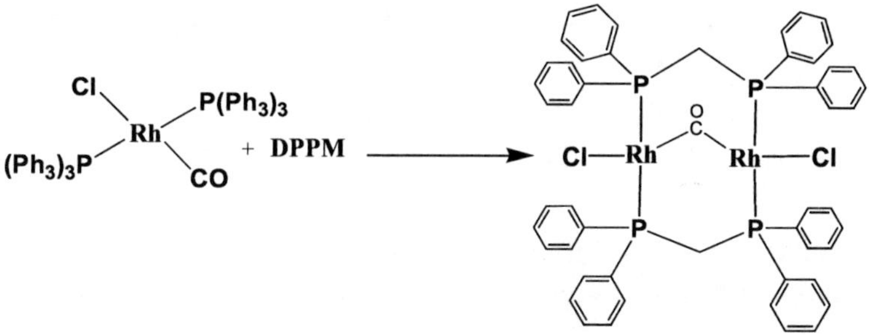

The accepted nomenclature notation for bridging ligands is the prefix descriptor η (eta) and for chelating ligands it is the prefix descriptor μ (mu). The prefix η^n indicates that a ligand is using n # of its atoms to form bonds to metal atoms. For example, the EDTA ligand when binding in a hexadentate fashion as shown on page 81 can be described as η^6-EDTA. For molecules with an obvious denticity, such as dppe or en, the eta descriptor is left off, as the denticity and # of ligating atoms is understood.

The prefix μ indicates that a ligand bridges only two metal atoms. If a ligand bridges three, or four metal atoms, the descriptors μ^3, or μ^4, and so on, are used.

Some unidentate ligands have two or more different donor sites so that the possibility of *linkage isomerism* arises. Some important ligands of this type, which are called *ambidentate ligands*, are:

$M-NO_2$ Nitro

$M-ONO$ Nitrito

$M-SCN$ Thiocyanato   or S-Thiocyanato

$M-NCS$ Isothiocyanato   or N-Thiocyanato

An example of two O- and N- bonded cobalt nitrito complexes are shown in the following scheme. Interestingly, the two forms are in equilibrium.

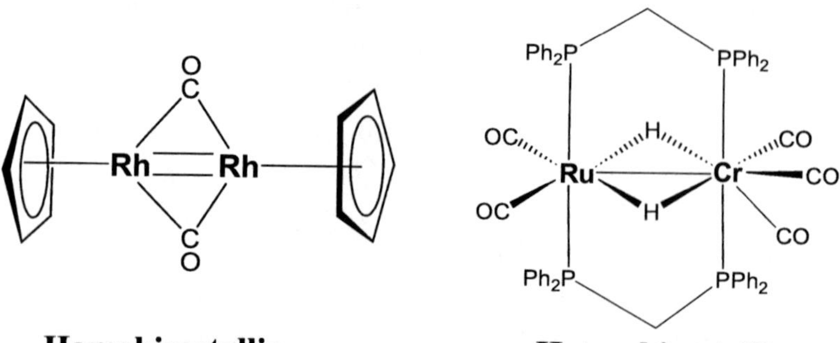

Heterobimetallic --------This term describes a complex in which there are two different metal centers, for example, ruthenium (Ru) and molybdenum (Cr).

Homobimetallic--------These complexes have two metal centers that are the same element. The two metals centers don't need to have identical ligands or coordination number, but are often found as symmetric dimers.

Ligands may also be ***macrocyclic*** in nature. These have even more thermodynamic stability than typical chelating ligands because of the rigid nature of the ligand itself.

An example is for the ***porphyrin*** ring shown below, and an example of a porphyrin analog binding to an atom of magnesium to form chlorophyll A. Other examples in nature find such porphyrin rings structures bound to Fe, Co, and Zn to form important biological enzymes.

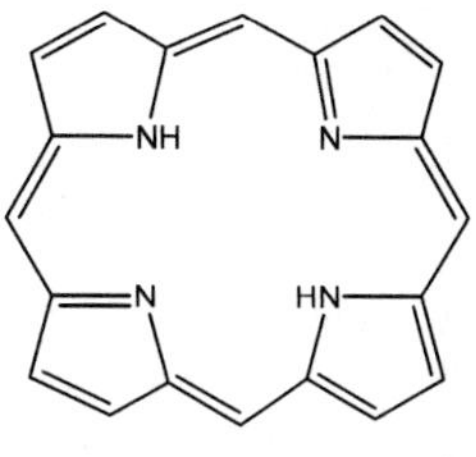

Porphyrin

Chlorophyll A

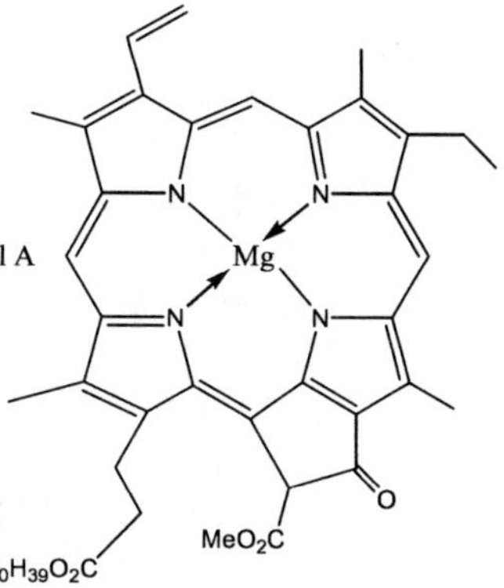

The **macrocyclic effect** is observed when a macrocyclic ligand replaces other ligands, including chelating ligands with the same denticity.

The macrocyclic effect is due to movement of the open arms in the open form compared to the lack of mobility for the closed macrocycle, which also gives more thermodynamic stability to the closed macrocycle.

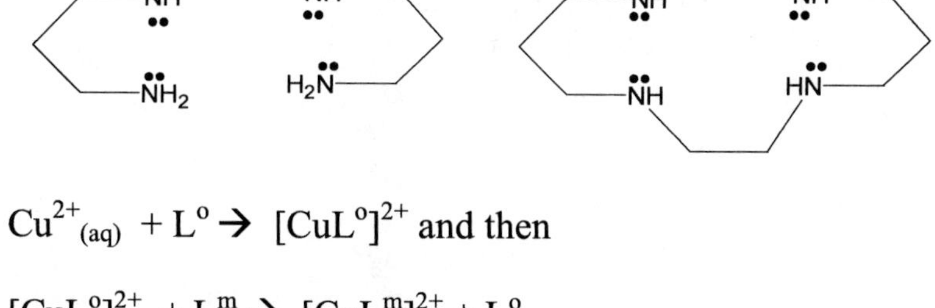

$$Cu^{2+}_{(aq)} + L^o \rightarrow [CuL^o]^{2+} \text{ and then}$$

$$[CuL^o]^{2+} + L^m \rightarrow [CuL^m]^{2+} + L^o$$

Thermodynamic Stability and Formation Constants

The thermodynamic stability of a coordination compound is often expressed by the equilibrium constant for the reaction of the aqueous metal ion, such as $[Co(H_2O)_6]^{3+}$, with the reacting ligand. For example:

$$[Co(H_2O)_6]^{3+} + 6\,NH_3 \leftrightarrow [Co(NH_3)_6]^{3+} + 6\,H_2O$$

This process proceeds by a stepwise displacement of all the water molecules with new ammonia ligands until all six have been replaced.

$(K_1)\ [Co(H_2O)_6]^{3+} + NH_3 \leftrightarrow [Co(H_2O)_5(NH_3)]^{3+} + H_2O$

$(K_2)\ [Co(H_2O)_5(NH_3)]^{3+} + NH_3 \leftrightarrow [Co(H_2O)_4(NH_3)_2]^{3+} + H_2O$

$(K_3)\ [Co(H_2O)_4(NH_3)_2]^{3+} + NH_3 \leftrightarrow [Co(H_2O)_3(NH_3)_3]^{3+} + H_2O$

$(K_4)\ [Co(H_2O)_3(NH_3)_3]^{3+} + NH_3 \leftrightarrow [Co(H_2O)_2(NH_3)_4]^{3+} + H_2O$

$(K_5)\ [Co(H_2O)_2(NH_3)_4]^{3+} + NH_3 \leftrightarrow [Co(H_2O)(NH_3)_5]^{3+} + H_2O$

$(K_6)\ [Co(H_2O)(NH_3)_5]^{3+} + NH_3 \leftrightarrow [Co(NH_3)_6]^{3+} + H_2O$

The overall equilibrium expression for this reaction, generally referred to as β_n (where n=6 for this reaction) is defined as:

$$\beta_6 = \frac{[Co(NH_3)_6]^{3+}}{[Co(H_2O)_6]^{3+}[NH_3]^6}$$

The **β_6** is the ***overall stability constant*** or the ***overall formation constant***. Water does not appear in the equilibrium expression. The values of K and β are related since β is the overall constant, therefore:

$\beta_6 = K_1 \times K_2 \times K_3 \times K_4 \times K_5 \times K_6$ and

$\log \beta_6 = \log K_1 + \log K_2 + \log K_3 + \log K_4 + \log K_5 + \log K_6.$

Table 3.1 of Common Ligands

__Common Name__	__Abbreviation__	
fluoro,	F^-	
chloro,	Cl^-	
bromo,	Br^-	
iodo,	I^-	
cyano,	CN^-	
thiocyano,	SCN^-	
isothiocyano,	NCS^-	
hydroxo,	OH^-	
aqua,	H_2O	
carbonyl,	CO	
thiocarbonyl,	CS	
nitrosyl,	NO^+	
nitro,	NO_2^-	
nitrito,	ONO^-	
oxalato,	$C_2O_4^{2-}$ or ox	
phosphine,	PR_3	
pyridine, C_5H_5N	pyr	
ammine, NH_3		
methylamine, CH_3NH_2		
ethylenediamine,	$NH_2CH_2CH_2NH_2$,	en
diethylenetriamine,	$NH_2C_2H_4NHC_2H_4NH_2$	dien
triethylenetetramine,	$NH_2C_2H_4NHC_2H_4NHC_2H_4NH_2$	trien
β, β', β''-triaminotriethylamine,	$N(C_2H_4NH_2)_3$	tren
acetylacetonato,	$[CH_3COCHCOCH_3]^-$	acac
2,2'-bipyridine,		bipy
1,10-phenanthroline,	$C_{12}H_8N_2$	phen
dialkyldithiocarbamate,	S_2CNR_2-	dtc
1,2-bis(diphenylphosphino)ethane,	$PPh_2C_2H_4,PPh_2$	dppe
ethylenediaminetetraacetate, $(-OOCCH_2)_2,NCH_2CH_2N(CH_2COO-)_2,$		EDTA

Structures and Abbreviations

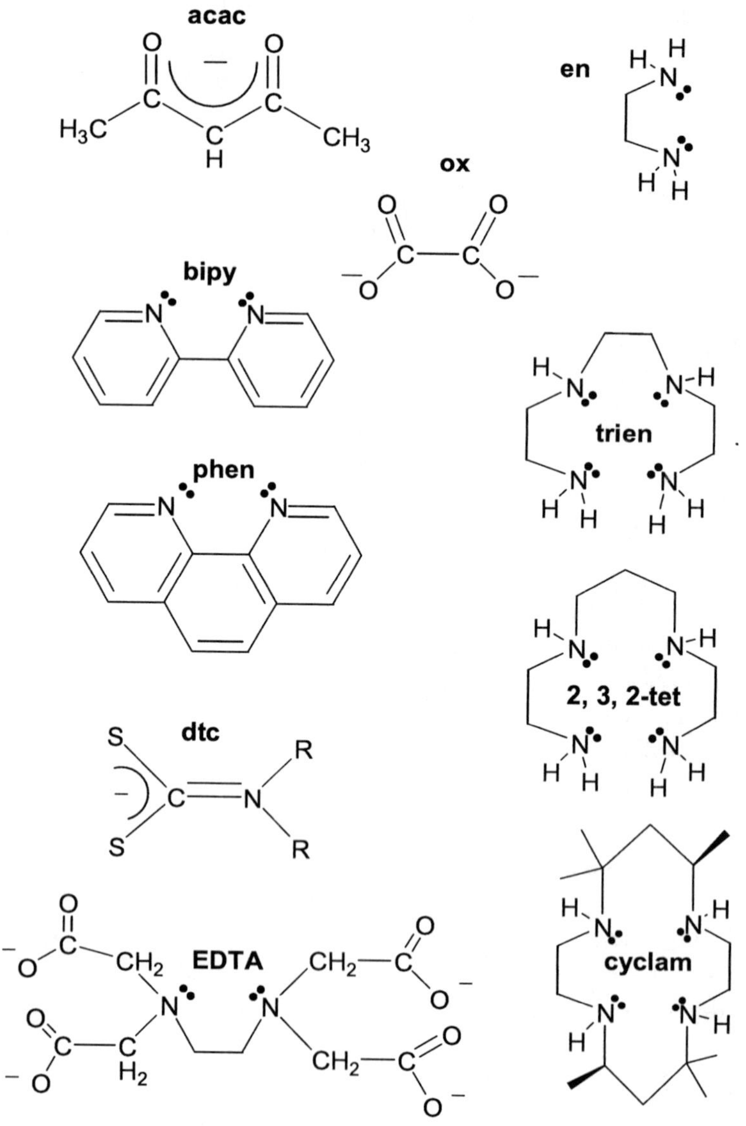

acac

en

ox

bipy

trien

phen

2, 3, 2-tet

dtc

EDTA

cyclam

Note: The ligands <u>acac, ox, dtc and EDTA</u> are shown as their deprotonated anions here in this table, not in their neutral protonated forms.

Monodentate Organophosphine Ligands

phosphine

trimethylphosphine

triphenylphosphine

trimethylphosphite

triphenylphosphite

tricyclohexylphosphine

trifluorophosphine

Bidentate Organophosphine Ligands

dmpm

dmpe

dppm

dppe

diphos

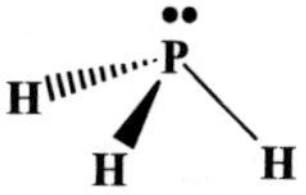

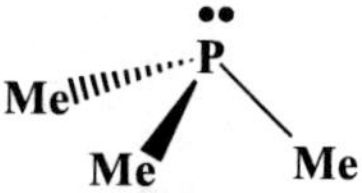

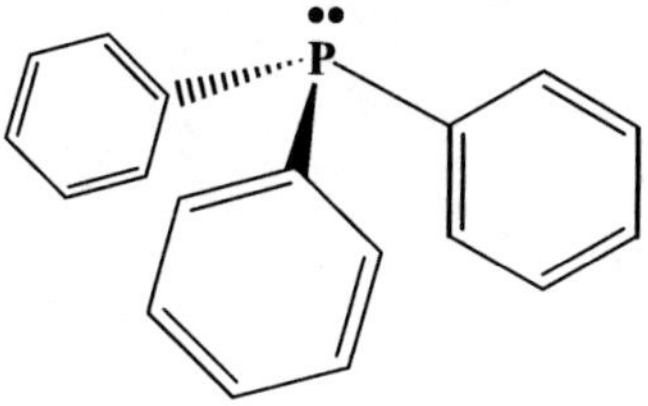

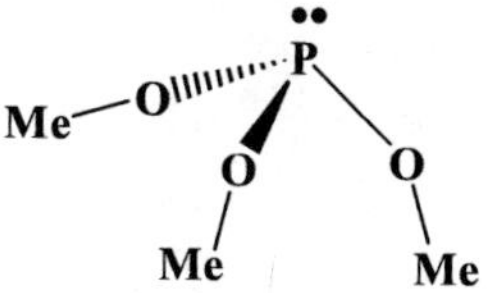

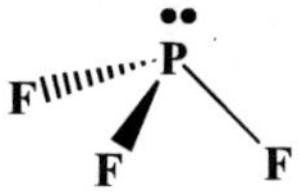

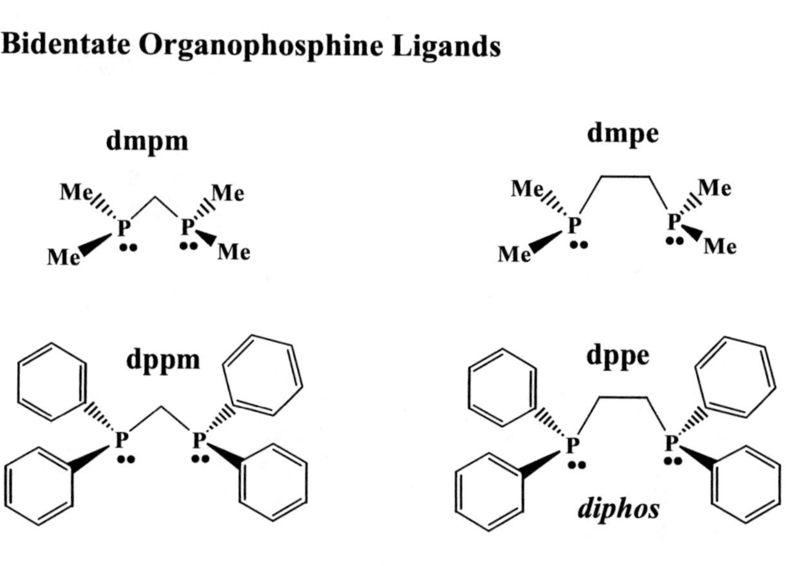

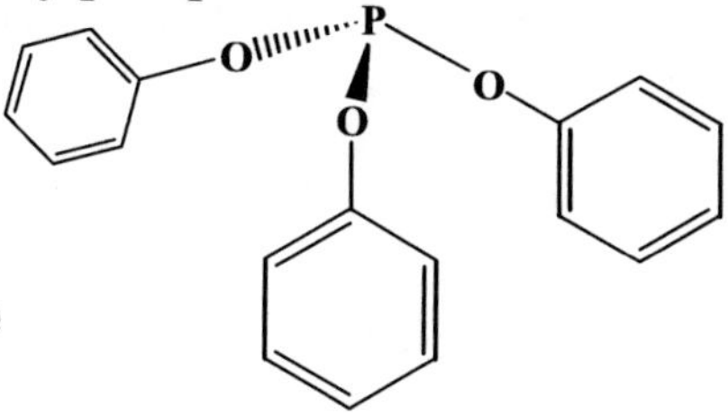

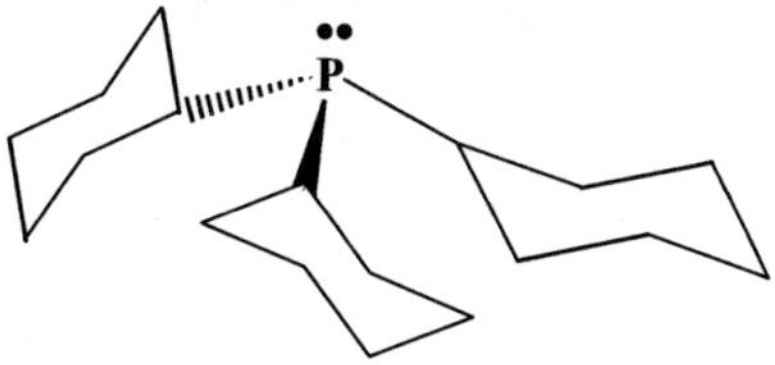

Chiral Organophosphine Ligands

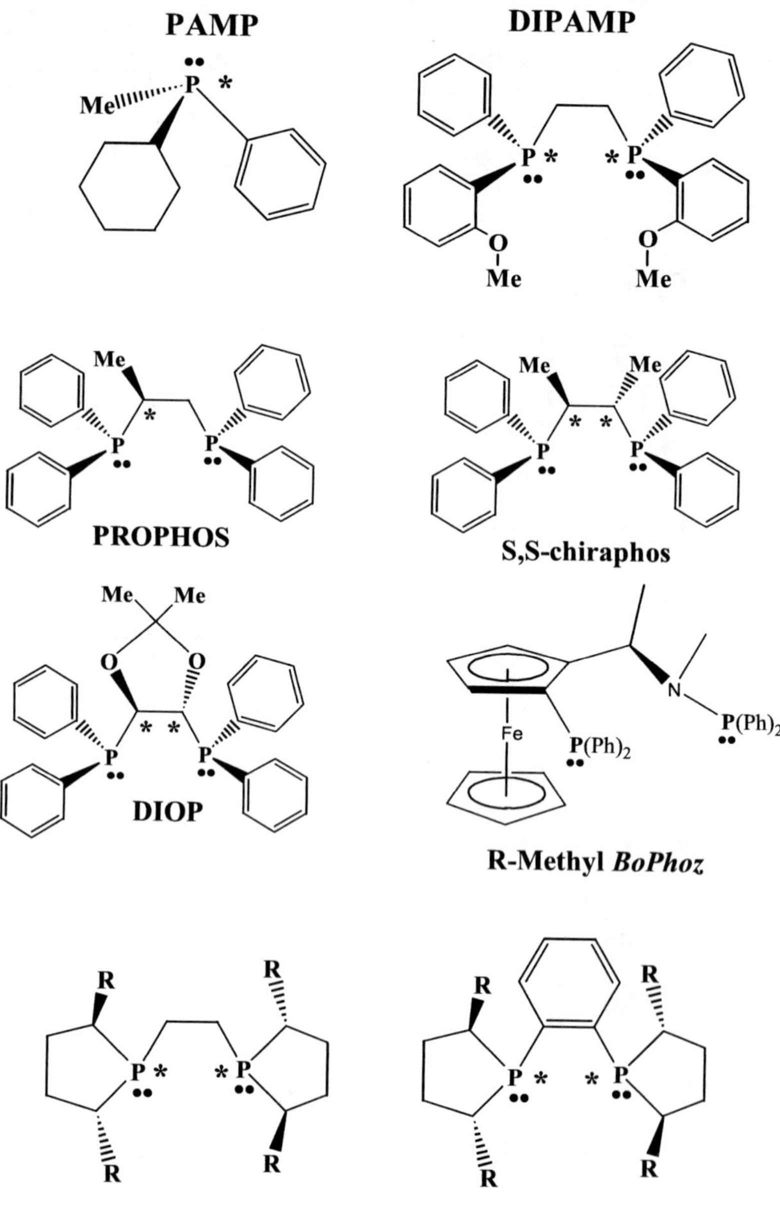

* indicates a chiral center

Nomenclature of Coordination Compounds

The nomenclature of simple coordination compounds is developed in a set of rules for referring to ionic and neutral ligands, the number of each type of ligand, and the oxidation state of the metal. A number of examples of naming compounds and writing formulas are given.

The nomenclature of coordination compounds is introduced in two sections. First, we consider the basics of <u>naming ligands</u> (including multidentate, ambidentate, and bridging) that occur in simple neutral as well as ionic coordination compounds, and then secondly we will consider <u>naming the coordination compounds</u> themselves.

Table 3.2 lists six rules for naming ligands and simple coordination compounds.

The name of anionic ligands is modified by removing the *-ide* suffix of halides, oxides, and hydroxides and replacing in these ligands with *-o*. Therefore, chloride becomes chloro, oxide becomes oxo, hydroxide becomes hydroxo, and so forth.

If the ligand has an ending such as *-ate* or *-ite*, the last *e* is removed and replaced with *-o*, so that nitrate becomes nitrato, sulfite becomes sulfite. But, there are a few exceptions to this rule. For example, amide becomes amido and the N-bonding form of the ambidentate nitrite becomes nitro.

The very few positive ligands that are found in coordination compounds are modified by adding an *–ium* suffix to the root name.

The names of neutral compounds that act as ligands are usually not modified, however, it is customary to give a few common neutral ligands special names. These special cases are: water becomes aqua, ammonia is called ammine, carbon monoxide is carbonyl, and nitrogen oxide is nitrosyl. Molecular nitrogen and oxygen are referred to as dinitrogen and dioxygen.

In naming a coordination compound, the cation is named first and the anion is named second. This is the same method used to name ordinary inorganic salts, like sodium bromide or ammonium phosphate.

Next, in naming a coordination compound, the ligands are always named first (in alphabetical order), followed by the name of the metal. (This can lead to ambiguity in naming compounds: you might ask which comes first---chloro or dibromo? The answer is that you use the name of the ligand in determining order---not the prefix denoting how many of them are present in the complex. Thus: di*bromo* and then *chloro* when naming.)

One should note that the opposite order is followed in writing the formulas for coordination compounds; the symbol for the metal precedes the formulas for the ligands. An example is the compound tetraamminedichlorocobalt(III) chloride, where the written formula is: $[Co(NH_3)_4Cl_2]Cl$.

As shown in the example above, the oxidation state of the metal is indicated by use of Roman numerals in parentheses after the name. Therefore an oxidation state of zero is indicated by a numeral zero, 0, in the parentheses, an oxidation state of +1 is (I), +2 is (II), +3 is (III), and so forth.

If the metal containing portion of the coordination compound is an anion, the *–ate* suffix is added to the name of the metal, but the *–ium* or other suffix has to be removed from the name of the metal before the *–ate* is added. For example, titanium becomes titanate, and the other metals such as manganese becomes manganate, and molybdenum becomes molybdenate. Some metals named with a Latin root, such as iron, copper, silver, and gold, will retain the Latin stem for the metal and become ferrate, cuprate, argentate, and aurate when they are anions.

The number of ligands is indicated by the appropriate prefix given in Table 3.2.

There are two sets of prefixes, one for monoatomic ions and polyatomic ions with relatively short names (*di-, tri-, tetra-,* etc.), and a second set of prefixes (*bis-, tris-, tetrakis-,* etc.) for ligands that already contain a prefix—for example, ethylenediamine or triphenylphosphine—or for complicated ligands whose names commonly appear in parentheses.

The use of parentheses is not as systematic in practice as might be expected, and is often used to help in clarification between different ligands, so that they won't be confused, and to help in setting off subscripts when the ligand itself has a subscript in the formula. Also, ionic ligands with particularly long names and neutral ligands without special names are enclosed in parentheses. So, for example, EDTA, or ethylenediammine-tetraacetato, is generally enclosed in parentheses, whereas cyano is not.

There are two different ways to describe ambidentate ligands. The first way is to put the symbol of the donating atom before the name of the ligand. This indicates which atom is coordinating to the metal, and is the preferred way to describe ambidentate ligands in that it infers the different ways the ligand could bind to the metal. Thus, –NCS would be N-thiocyanato, and –SCN would be called S-thiocyanato. The second way is to use a slightly different form of the name, which is dependant on the atom that is donating the electron pair to the metal. Thus, –NCS would be isothiocyanato, and –SCN would be called thiocyanato.

Bridging ligands are named by placing the Greek letter μ (mu) before the name of the ligand.

Therefore, a bridging amide (NH_2^-), a bridging hydroxide (OH^-), or a bridging sulfide (S^{2-}) ligand becomes μ-amido, μ-hydroxo, or μ-sulfido, respectively.

If there is more than one of a given bridging ligand, the prefix indicating the number of ligands is placed after the μ and before the ligand. Thus, if there are two bridging hydroxo ligands they are indicated by using *μ-dihydroxo*.

If there are two bridging ligands, one bromide and one chloride for example, they are given in alphabetical order as *μ-bromo, μ-chloro*.

Table 3.2. Nomenclature Rules for Simple Ligands

1. Anionic ligands end in *–o*.

F^- fluoro	SO_3^{2-}	sulfito	OH^-		hydroxo
Cl^- chloro	SO_4^{2-}	sulfato	CN^-		cyano
Br^- bromo	$S_2O_3^{2-}$	thiosulfato	NC^-		isocyano
I^- iodo	ClO_3^-	chlorato	SCN^-		thiocyanato
O^{2-} oxo	CH_3COO^- acetato		NCS^-		isothiocyanato

2. Neutral ligands are named as the neutral molecule.

C_2H_4	ethylene
$(C_6H_5)_3P$	triphenylphosphine
$NH_2CH_2CH_2NH_2$	ethylenediamine
CH_3NH_2	methylamine

3. There are special names for four neutral ligands:

H_2O is *aqua*, NH_3 is *ammine*, CO is *carbonyl*, NO is *nitrosyl*.

4. Cationic ligands end in *–ium*.

$NH_2NH_3^+$ hydrazinium

5. Ambidentate ligands are indicated by:
   a. Placing the symbol of the coordinating atom in front of the name of the ligand, for example, *S*-thiocyanato and *N*-thiocyanato for –SCN- and –NCS

   b. Using special names for the two forms, for example, nitro and nitrito for $–NO_2^-$.

6. Bridging ligands are indicated by placing a μ-before the name of the ligand.

Table 3.3. Naming Simple Coordination Compounds

1. As in simple salts, the name of the cation is given first.

2. The contents of the coordination sphere are specified: first the ligands, then the metal.

(a) The nature of the ligands is indicated by a suffix: positively charged ligands have the suffix *-ium*, neutral ligands no suffix, and negative ligands usually have the suffix *-o* (but H_2O and NH_3 are called *aquo* or *aqua* and *ammine*, respectively). (**See Table 3.2**)

(b) The ligands are named in a definite order according to their charge: anionic ones first, then neutral, then cationic. Conventionally, but not necessarily conveniently, the names are not separated by spaces or hyphens.
(**See Table 3.2**)

(c) The number (two, three, four) of each kind of ligand is indicated by prefixes, *di, tri, tetra-* (for ligands having simple names such as chloro), or *bis-, tris, or tetrakis-* (for ligands having complicated names such as 2,4-pentanediono). (**See Table 3.4**)

(d) Finally, the name of the metal is given, the oxidation state is indicated by a Roman numeral in parentheses, and the ionic nature of the complex is indicated by the suffix *-ate* for an anionic complex or no suffix for cationic or neutral complexes.

Table 3.4. EXAMPLES OF NOMENCLATURE OF COORDINATION COMPOUNDS

$[PtCl_2(NH_3)_4]Br_2$ = Dichlorotetraammineplatinum (IV) bromide
$[Al(OH)_2(H_2O)_4]^+$ = Dihydroxotetraaquoaluminum (III) ion
$[Co(NO_2)(NH_3)_5]SO_4$ = Nitropentaamminecobalt (III) sulfate
$[Co(ONO)(NH_3)_5]SO_4$ = Nitritopentaamminecobalt (III) sulfate
$[Co(H_2NCH_2CH_2NH_2)_3]Cl_3$ =
Tris(ethylenediamine) cobalt (III) chloride
$NH_4[Cr(NCS)_4(NH_3)_2]$ =
Ammonium tetra-n-thiocyanatodiammine-chromate (III)
$Na_3[Fe(CN)_6]$ = Sodium hexacyanoferrate (III)
$Na_3[Co(NO_2)_6]$ = Sodium hexanitrocobaltate (III)
$[Zr(acac)_4]$ = Tetrakis(2,4-pentanediono) zirconium (IV) or
tetrakisacetylacetonato zirconium(IV)

Examples:

The best way to understand Table 3.1, Table 3.2, Table 3.3, Table 3.4, and the text explanation is by working through a series of examples.

The rules for writing these formulas, as for all chemical compounds, are determined by the International Union of Pure and Applied Chemistry (IUPAC). We first consider naming compounds for which we are given a formula.

Example A
Name the compound $[Co(NH_3)_4Br_2]Br$.

One starts first by naming the complex cation. The ligands are named alphabetically with ammine first and then bromo, and since there are four ammonias and two bromides the prefixes *tetra-* and *di-* are used. The cobalt oxidation state is determined as follows: The net charge on the complex cation must be 1^+ to balance the one 1^- bromide anion. Since there are two 1^- bromides in the coordination sphere, the cobalt must be +3 oxidation state in order for the net charge on the cation to come out as 1^+. Thus, the full name of the compound is:

tetraamminedibromocobalt(III) chloride
or   dibromotetraamminecobalt(III) chloride

Example B

Name the compound $(NH_4)_2[Pd(SCN)_6]$.

Here is a palladium-containing complex (which is an anion) and the counter ion is the common ammonium ion, NH_4^+, as the cation. Since the ligand is written with the S symbol first, it is the thiocyanato (or, alternatively, *S*-thiocyanato) form of the ambidentate ligand. There are six of these thiocyanate ligands, so one must use the *hexa*-prefix. The metal complex anion must have a net charge of 2^- to balance the two 1^+ ammonium cations. Since the thiocyanate ion is also 1^-, the palladium must be in the +4 oxidation state to give a net 2^- charge to the anion. Because the palladium is contained in a complex anion, its *–ium* suffix is removed and replaced with *–ate*. Therefore the full name of the compound is:

ammonium hexathiocyanatopalladate(IV)

Example C

Name the compound $[Zn(en)_2]SO_4$.

The zinc complex is a complex cation since there is a sulfate counter ion. There are two ethylenediamine ligands, which are shortened to *en*, but since en is a neutral ligand with *di-* already within its name, the prefix *bis-* is used. The oxidation state of the zinc will be the same as the charge on the overall complex cation since the ligands are neutral. The charge on the metal complex has to be 2^+ to balance the 2^- of the sulfate ion, so the zinc is in the +2 oxidation state. The full name of this compound is therefore:

bis(ethylenediamine)zinc(II) sulfate

Example D
Name the compound $[Ag(CH_3NH_2)_2]_2SO_4$.

One must use the *bis-* prefix for the two neutral methylamine ligands. The silver metal complex is a cation, and there are two of them so as to balance out the 2^- charge on the sulfate ion. Thus, the oxidation state of the silver has to be +1. The full name of this compound is therefore:

bis(methylamine)silver(I) sulfate

Example E
Name the compound $N(butyl)_4[Re(H_2O)_2(C_2O_4)_2]$.

The tetrabutyl ammonium ion, and all ammonium type ions, are 1^+ charged, thus the rhenium metal complex is an anion with a 1^- charge. In the metal complex anion, the two neutral aqua (water) ligands come alphabetically before the two anionic oxalate ligands. The oxalate ligands are each 2^- charged. Since the two oxalate anionic ligands contribute 4^- charhes, and the overall complex has a 1^- charge, then the rhenium is in the +3 oxidation state. The name rhenium is amended to rhenate because this metal is in a complex anion. Therefore the full name of this compound is:
Tetrabutylammonium diaquadioxalatorhenate(III)

Example F
Name this compound:

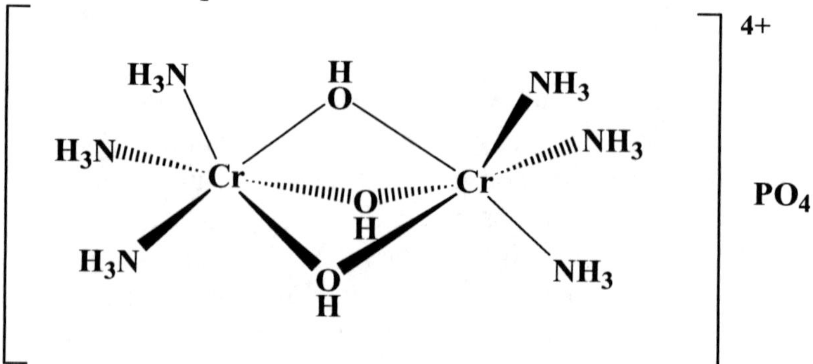

This is an example of a bridged compound. The three hydroxide ligands bridge between the two chromium ions.

One names such compounds from left to right remembering to place a μ in front of the bridging ligands. The oxidation states of the two chromium metal atoms could be (III) and (III), (II) and (IV), (I) and (V), or any other combination adding up to 6. If one were to find out that the two metal atoms have the same oxidation state, then the oxidation state of chromium would be +3. The full name of the compound is therefore:

triamminechromium (III)- μ–trihydroxo triamminechromium (III) phosphate

Instead of writing the formula from the structure, it is instructional to work a few examples in which one is given the names of some coordination compounds and then to supply the correct formulas or structures.

The IUPAC rules concerning the order in which the formulas of the ligands in a coordination compound should be written are complicated and generally not treated in an introductory textbook. We will follow the common, but not officially correct method, practice of writing the ligand formulas of a coordination compound in the same order they are named, which is in alphabetical order by the first letter of the name of the ligand.

Example G
Write the formula for the compound bis(acetylacetonato)diaquarhodium(III) chloride.

The formula for the bidentate acetylacetonate ligand is given in Table 3.1, but this 1^- anion is usually abbreviated as **acac**. The acac and two water molecules that constitute the coordination sphere along with the rhodium(III) cation are set apart in brackets. The net charge on the complex cation is 1^+ (because the acac is 1^- and there are two of them); so one chloride counter anion is needed. The formula of the compound is therefore:

$$[Rh(acac)_2(H_2O)_2]Cl$$

Example H

Write the formula for the compound diamminebromo(ethylene)nitrotriphenylphosphineplatinum (IV) phosphate.

This compound has five different types of ligands in the coordination sphere: NH_3, Br^-, C_2H_4, NO_2^-, and PPh_3. The real difficulty in constructing this formula is figuring out how many cations and anions there must be. The cation has a net charge of 2^+, and the anion is 3^-. Therefore, there must be three cations and two anions to ensure electrical neutrality. The formula for this compound is therefore:

$$[Pt(NH_3)_2Br(C_2H_4)NO_2PPh_3]_3(PO_4)_2$$

Example I

Draw the structural formula for:
 tetraamminemolybdenum (III)- μ-chloro- μ-hydroxo bis (ethylenediamine)ruthenium(III) sulfate.

The OH^- and Cl^- ions are bridging ligands between the cobalt and iron cations. The overall charge on this huge cation is 4^+ because there are 6^+ charges from the two 3^+ cations and 2^- charges from the two 1^- bridging hydroxo and chloro anions. Therefore, there must be two (2^-) sulfates in the formula:

Terms and Definitions Chapter 3:

Ligand
Hard and Soft
Polarizable
Nucleophiles
Electrophiles
Coordinate Covalent Bond
Classical Or Simple Donor Ligands
Nonclassical Ligands
π-Bonding or π-Acid –Ligands
One-Electron Donors
Two-Electron Donor
Three-Electron Donors

Unidentate
Bidentate

Chelate

Bridging Groups

Bidentate Bridging
Multidentate Ligands

Inherent Thermodynamic Stability

Linkage Isomerism
Ambidentate Ligands

Heterobimetallic
Homobimetallic

Macrocyclic
Overall Formation Constant
Overall Stability Constant

Chapter 3 Problems and Exercises

1. In the compound Al_2Cl_6, two of the chloro ligands are *bridging* and four are terminal. Draw this structure.

2. Describe whether the following ligands are monodentate or bidentate:
(A) ethylenediamine
(B) aqua
(C) cyano
(D) carbonyl
(E) ammine

3. In Lewis acid base theory, a d-metal complex consists of a central metal atom to which are attached a number of ligands.   The metal is a ________________ and the ligands are ________________ . (Lewis Acid or a Lewis Base?)

4. Which of the following ligands are considered to be one-electron donors?
    (a) chloride (b) ammonia (c) water (d) CH_3^- (e) iodide

5. Is the Cu^{2+} ion considered to be a Lewis acid, or a Lewis base?

6. Which of the following ligands in each listed pair is considered to be the softest?
    (a) F^- or I^-
    (b) S^{2-} or O^{2-}
    (c) PH_3 or AsH_3
    (d) en or dppe
    (e) thiocyano or isothiocyano
    (f) EDTA or dtc (dithiocarbamate)
    (g) ammonia or pamp

7. Which metal in each pair listed is the hardest?
    (a) Cu^+ or Cu^{2+}
    (b) Pd^{2+} or Ni^{2+}
    (c) Cr^{3+} or Au^{3+}

8. Which of the following ligands in each pair listed would *most likely* have the greatest inherent stability in binding to a metal ion?

   (a) Trimethyl phosphite or diphos
   (b) ammonia or EDTA
   (c) acac or water
   (d) chloride or diop
   (e) en or porphyrin
   (f) dmpe or s,s-chiraphos
   (g) trien or cyclam

9. Label each of the following ligands as simple classical ligands or non-classical ligands.

   (a) ammonia (b) water (c) chloride (d) EDTA (e) en
   (f) diphos (g) acac (h) DuPhos (i) porphyrin
   (j) dipamp (k) carbon monoxide (l) methylamine

10. Draw the structures of the two compounds formed when dmpe and dmpm both are used as a chelating bidentate ligand in separate reactions with chromium hexacarbonyl, $Cr(CO)_6$. What is the difference in the chelate rings formed by these two ligands in the reaction with $Cr(CO)_6$

11. Name the following compounds:
(a) $[Pt(NH_3)_4Cl_2]SO_4$
(b) $K_3[Mo(CN)_6F_2]$
(c) $K[Co(EDTA)]$
(d) $[Co(NH_3)_3(NO_2)_3]$

12. Name the following compounds:
(a) $[Pt(NH_3)_6]Cl_4$
(b) $[Ni(acac)(P(C_6H_5)_3)_4]NO_3$
(c) $(NH_4)_4[Fe(ox)_3]$
(d) $W(CO)_3(NO)_2$

13. Name the following compounds:
(a) $[Pt\{P(C_6H_5)_3\}_4](CH_3COO)_4$
(b) $Ca_3[Ag(S_2O_3)_2]_2$
(c) $Ru(As(C_6H_5)_3)_3Br_2$

14. Name the following compounds:
(a) $[Fe(en)_3][IrCl_6]$
(b) $[Ag(NH_3)(CH_3NH_2)]_2[PtCl_2(ONO)_2]$
(c) $[VCl_2(en)_2]_4[Fe(CN)_6]$

15. Write formulas for the following compounds:
(a) Pentaammine(dinitrogen)ruthenium(II) chloride
(b) Aquabis(ethylenediamine)thiocyanatocobalt(III) nitrate
(c) Sodium hexaisocyanochromate(III)

16. Write formulas for the following compounds:
(a) Bis(methylamine)silver(I) acetate
(b) Barium dibromodioxalatocobaltate(III)
(c) Carbonyltris(triphenylphosphine)nickel(0)

17. Write formulas for the following compounds:
(a) Tetrakis(pyridine)bis(triphenylarsine)cobalt(III) chloride
(b) Ammonium dicarbonylnitrosylcobaltate(-I).
(c) Potassium octacyanomolybdenate(V)
(d) Diamminedichloroplatinum(II)

18. Write formulas for the following compounds:
(a) Hexamminecobalt(III) pentachlorocuprate(II)
(b) Tetrakis(pyridine)platinum(II) tetrachloroplatinate(II)
(c) Diammine bis(triphenylphosphine) palladium(II)
(d) bis(oxalato) aurate(III)

19. Draw a valid structure for each of the following
compounds:
(a) (Ethylenediamine)iodonitritochromium(III)- μ-
dihydroxotriamminechlorocobalt(III) sulfate
(b) Bis(ethylenediamine) cobalt(III)- μ-isocyano- μ-
thiocyanatobis(acetylacetonato) chromium(III)nitrate

20. Draw a valid structure for each of the following compounds:
(a) Pentamminechromium(III)- μ-hydroxopentamminechromium(III) chloride
(b) Diammine(ethylenediamine)chromium(III)- μ-bis(dioxygen)tetraaminecobalt(III) bromide.

21. The ligand thiocyanate (SCN-) is an ambidentate ligand. Draw two valid tetrahedral metal complexes of Zn^{2+} of this ligand that each exhibit one of the ambidentate bonding modes that this ligand can use to bind to the metal.

22. Which of the following bidentate chelating ligands would probably exhibit the most steric hindrance around a metal ion?
    (a) dmpm or dmpe
    (b) dppe or dppm
    (c) dmpe or dppe
    (d) dppe or dipamp
    (e) triphenyl phosphine or triphenyl phosphite
    (f) trien or 2,3,2-tet

23. Label the following compounds as homobimetallic or heterobimetallic.
    (a) $Co_4(CO)_{12}$ (b) $Co_3Rh(CO)_{12}$
    (c) $Mn_2(CO)_{10}$ (d) $Fe_2Ru(CO)_{12}$

24. The total displacement reaction of water with ammonia ligand, on aqueous cobalt 3+ is given as:

$$[Co(H_2O)_6]^{3+} + 6\,NH_3 \leftrightarrow [Co(NH_3)_6]^{3+} + 6\,H_2O$$

    If the stepwise stability constants for the reaction at a certain temperature and pH is:
Log $K_1 = 2.79$, Log $K_2 = 2.26$, Log $K_3 = 1.69$, Log $K_4 = 1.25$, Log $K_5 = 0.74$, Log $K_6 = 0.03$.

    (a) calculate log β_6 for $[Co(NH_3)_6]^{3+}$
    (b) calculate β_6

Chapter 4.   Stereochemistry of Coordination Compounds

Coordination compounds exhibit a wide variety of isomers due to large coordination numbers (usually 4-6) and the differing ways in which ligands can bind to the metal center.

There are many types of isomers in coordination chemistry, for example: ***solvent isomers, ionization isomers, coordination isomers***; all of these have different ligands in the coordination sphere, with the same overall formula. The names of these ***constitutional isomers*** indicate whether solvent, anions or other coordination compounds form the changeable part of the structure.

Stereoisomers are isomers with the same ligands but with different shapes. These ***configurational isomers*** are the main focus of this chapter.

The following diagram may help make the distinction between isomer types clearer.

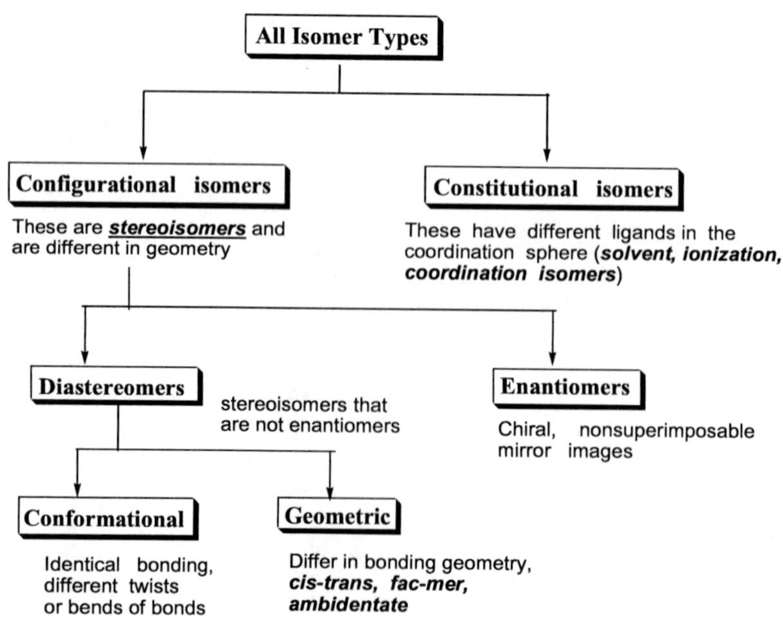

The terms *cis and trans* are common in organic chemistry and have approximately the same meaning for coordination compounds. However, the terms *fac* and *mer* are unique to coordination chemistry, and like the terms *cis* and *trans*, they describe different bonding geometries of ligands around the metal center. The terms *linkage isomerism* or *ambidentate isomerism* are used for cases of bonding through different atoms of the same ligand as discussed in the previous chapter.

Configurational Isomers

Examples of typical stereoisomerism are discussed next for four coordinate metal complexes (square planar) and six coordinate metal complexes (octahedral). Four coordinate geometries common for transition metals are square planar and tetrahedral. The tetrahedral geometry is so thoroughly covered in organic classes, with so many examples, that it is really not appropriate to cover that material in this text. On the other hand, square planar and octahedral geometries are not covered in organic classes.

Square Planar

Square planar complexes may exhibit geometrical isomers when there is more than one ligand. There can be up to four separate ligands on a metal with a square planar geometry. If A, B, C, and D are the four different ligands then we can have the following metal compounds with geometrical isomers: (MA_2B_2), (MA_2BC), and ($MABCD$).

The MA_3B compound (or any of its analogs with ligands C or D) would only have one isomer.

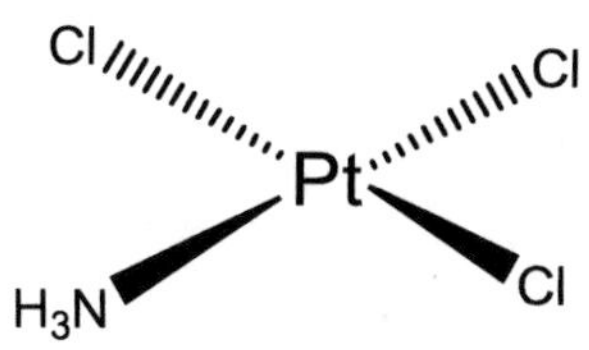

The examples for the three cases are discussed below.

4-coordinate square planar (MA_2B_2) = two isomers
Cis and trans isomers of $[PtCl_2(NH_3)_2]$

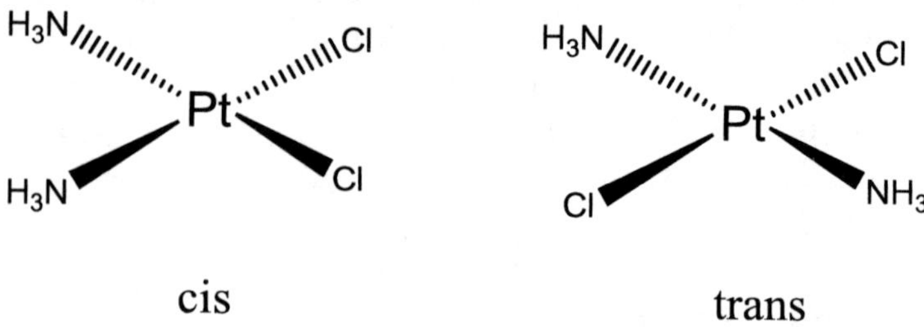

cis   trans

4-coordinate square planar (MA_2BC) = two isomers
Just as in the example above, there can only be two isomers:

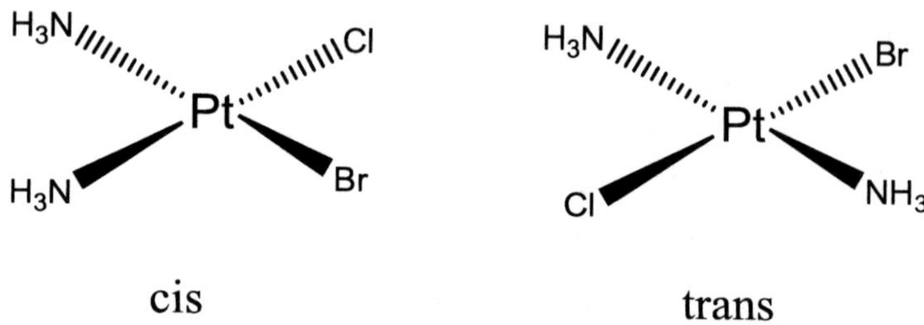

cis   trans

4-coordinate square planar (MABCD) = three isomers
There are three isomers, and the key to seeing them is to notice what ligands are *trans* to each other. (Ammonia is trans to a different ligand in each of the three isomers)

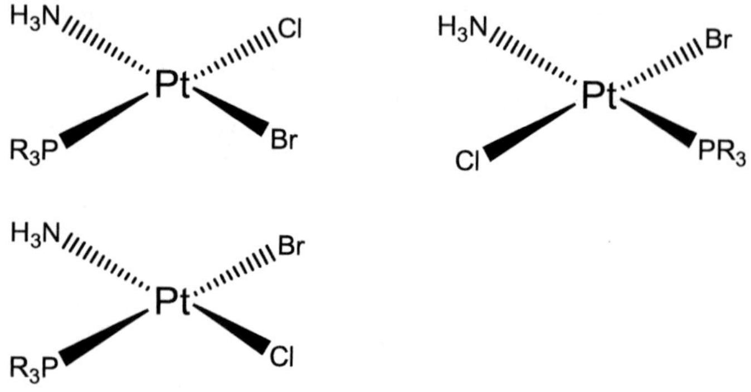

Octahedral

The octahedral geometry is by its very nature much more complex than the square planar geometry when it comes to isomer determinations. It is best to look at the simplest cases first.

The simplest case of isomer formation for an octahedral complex is when there is only one kind of ligand, such as ethylene diamine, a bidentate chelating ligand. For the case where there are three bidentate chelating ligands around the metal center, such as in $[Co(en)_3]^{3+}$, there is the possibility of two *enantiomers* (enantiomers are isomers that are ***non-superimposable mirror images***).

These two stereoisomers are shown in the next scheme. They can be thought of as propeller blades to give a better feel for their structure. These are ***mirror images***.

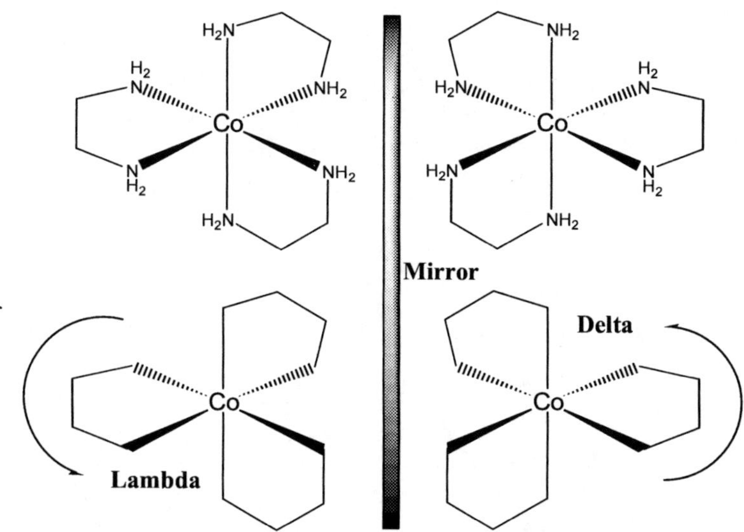

Molecules like these that are *nonsuperimposable mirror images* of each other are said to be ***chiral***.

To decide if a particular isomer has a Lambda or Delta configuration one must look down the 3-fold rotation axis and see if the propeller rotates counterclockwise (Lambda, Δ) or clockwise (Delta, Λ).

For an octahedral complex with two different ligands there can be various types of ligands, such as monodentate, bidentate, or even tridentate. With a tridentate ligand such as dien, in the following case **[M(L-L-L)X$_3$]**, two geometric isomers are possible:

[Co(dien)Cl$_3$] with a ***tridentate chelating ligand***

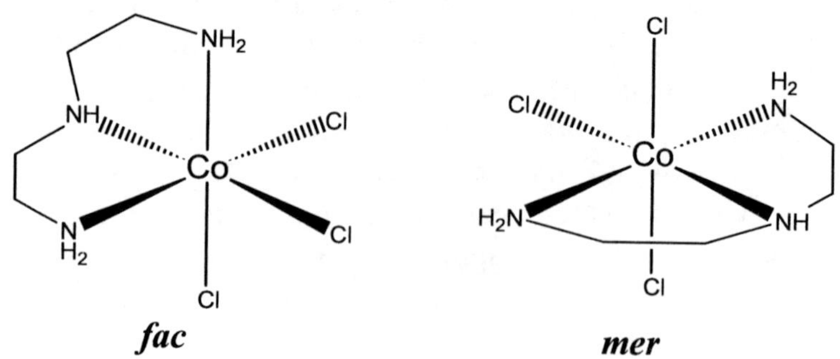

The fac isomer (facial) is named in that manner because it occupies one face of the octahedron, and the mer isomer (meridional) is named in that manner because it occupies the meridian of the octahedron (lies in a plane).

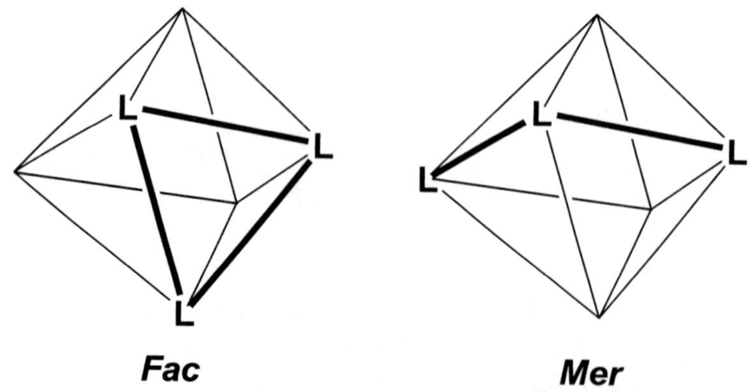

The situation on the previous page, where both fac and mer isomers are seen, is similar to the case where there are two monodentate ligands coordinated to the metal in equal numbers; only these same two geometric isomers are observed:

[MA₃B₃] case such as [Co(NH₃)₃Cl₃]

fac (facial)   **_mer_** (meridional)

Compare these fac and mer isomers to the ones on the preceding page.

With varying numbers of two different ligands specific isomer cases are observed. For example, only two geometric isomers are observed in the [MA₄B₂] case:

[MA₄B₂] case such as [Co(NH₃)₄Cl₂]⁺

cis   trans

Cis and trans isomers are seen in this case. Compare this set of isomers to the square planar case that was discussed previously. The similarity is that the ligands around the square plane determine the cis-trans isomer observed.

For monodentate ligands the following can be summarized:
[MA$_5$B$_1$] case = one isomer only
[MA$_2$B$_4$] case = two isomers (cis and trans)
[MA$_3$B$_3$] case = two isomers (fac and mer)
[MA$_4$B$_2$] case = two isomers (cis and trans)
[MA$_1$B$_5$] case = one isomer only

A more complicated situation arises when there are two bidentate ligands and two monodentate ligands around the octahedral coordination sphere. In this case, not only do we observe geometric isomers (cis/trans), but one of the isomers also has an enantiomer (a non-superimposable mirror image).

An example is: **[M(L-L)$_2$X$_2$] case** such as [Co(en)$_2$Cl$_2$]$^+$

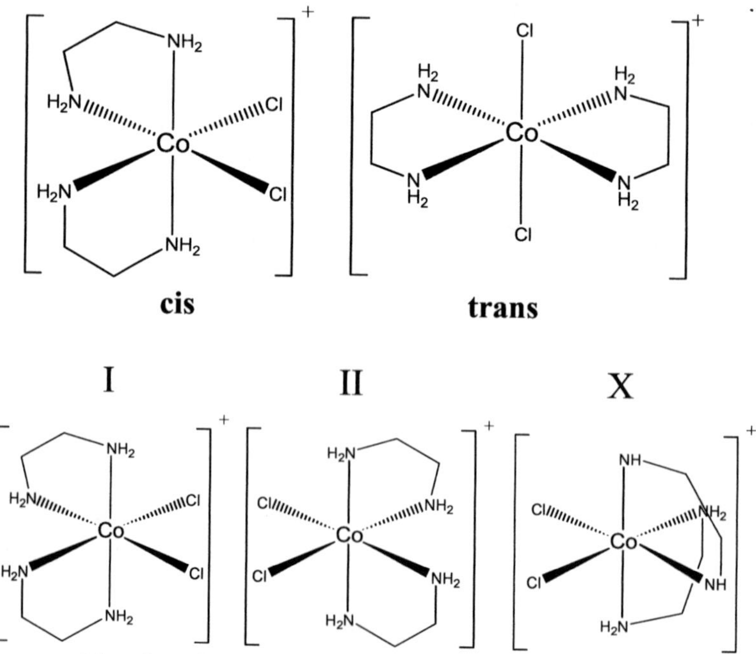

For the cis geometric isomer, there are two enantiomers, these are mirror images I and II. Isomer II is not the same as structure X, which is shown so that a comparison can be made. Structures I and X are the same after simple rotation by 180° around the cobalt.

[MA₂B₂C₂] case has six isomers (five are geometric)

The case where there are three different ligands, two of each, can be worked out systematically.

Ligands of the same type can either be cis or trans to each other in an octahedral complex. All three of the ligands can express a cis orientation, but only one ligand type at a time can be trans. If two ligand types are trans to each other, then all three must be trans. So, the following geometric isomers can be drawn for the $[Co(NH_3)_2(Br)_2(Cl)_2]^-$ ion:
(1) trans, trans, trans.
(2) trans, cis, cis (3) cis, trans, cis (4) cis, cis, trans
(5) cis, cis, cis

Thus, we have an all cis isomer, an all trans isomer, and three isomers that have two ligand types cis, and the other trans. The all cis isomer has an enantiomer. We can tell that the all cis isomer would have an enantiomer since it has no plane of symmetry through it, whereas we can draw a plane of symmetry through the other geometric isomers.

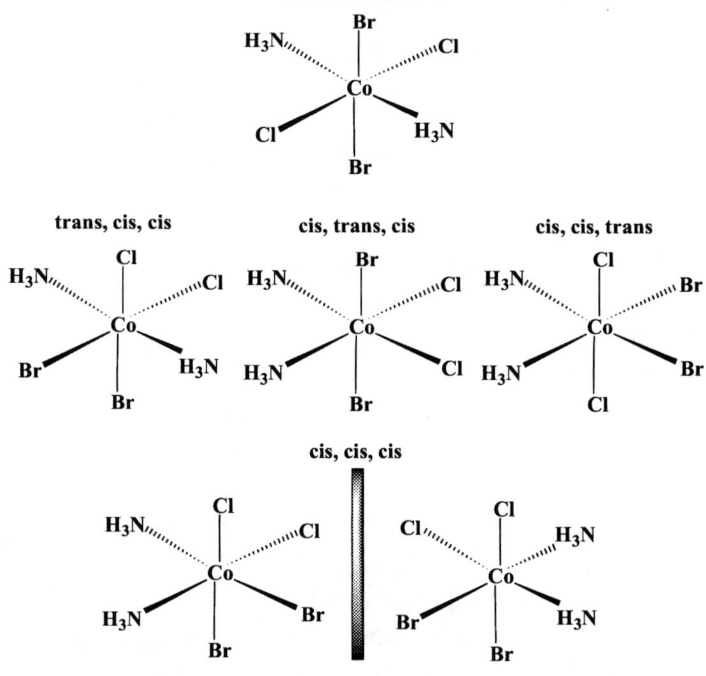

For octahedral complexes with three or more different ligands the situation can become very complicated as far as total number of isomers, enantiomers, and equivalent structures. The best way to understand these is to work the exercises at the end of this chapter. These exercises use some real examples of compounds that exhibit numerous isomers.

Constitutional Isomers

In octahedral metal complexes the six octahedral positions are commonly numbered as below, with positions 1 and 6 in axial positions and 2-5 in counterclockwise order as viewed from the 1 position.

If six different monodentate ligands are completely scrambled (rather than limited to trans pairs as shown on the previous page), then there are 15 different diastereomers (different structures that are not mirror images of each other), each of which has an enantiomer (mirror 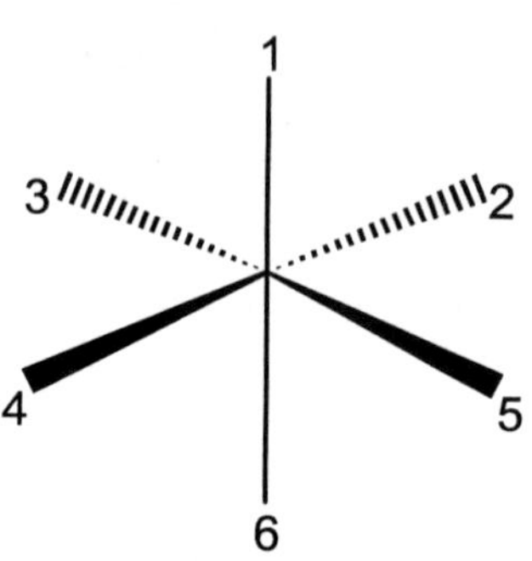image), which means there can be 30 different isomers! There is only one example of an octahedral Pt compound to date (that we are aware of) that has six different ligands around the metal.

Hydrate isomerism is also called ***solvent isomerism***.

The classical example is $CrCl_3$ x 6 H_2O. This compound can have three distinctive forms now known as:

$[Cr(H_2O)_6]Cl_3$	violet
$[CrCl(H_2O)_5]Cl_2$ x H_2O	blue-green
$[CrCl_2(H_2O)_4]Cl$ x 2 H_2O	dark green

Draw these compounds, and which, if any, exhibit isomerism?   Could you separate the three compounds above by using an ion exchange column? **Yes!**

What kind of resin would you use? Would you use an anion exchange resin, or a cation exchange resin?

Ionization Isomerism is exhibited by compounds with the same formula, but which give different ions in solution. Examples are:

$$[Co(NH_3)_5SO_4]NO_3 \quad \text{and} \quad [Co(NH_3)_5NO_3]SO_4$$

$$[Co(NH_3)_4NO_2Cl]Cl \quad \text{and} \quad [Co(NH_3)_4Cl_2]NO_2$$

Coordination Isomerism *requires at least two metal ions.*
An example is the set of platinum compounds with the ***empirical formula*** of $[Pt(NH_3)_2Cl_2]$
There are three coordination isomers possible:
- $[Pt(NH_3)_4Cl][PtCl_4]$
- $[Pt(NH_3)_2Cl_2]$
- $[Pt(NH_3)_3Cl] [Pt(NH_3)Cl_3]$

Linkage Isomerism:
Some ligands can bond to the metal through different atoms, and these are called ***ambidentate ligands***.

Examples: thiocyanate SCN-   can bind through S or N
nitrite NO_2-   can bind through N or O

Terms and Definitions for Chapter 4

Solvent Isomers
Ionization Isomers
Coordination Isomers
Constitutional Isomers
Stereoisomers
Chiral
Configurational Isomers
Hydrate Isomerism
Solvent Isomerism
Coordination Isomerism
Empirical Formula

Chapter 4 Problems and Exercises

1. The octahedral compound $[Co(NH_3)_3Br_3]$ exhibits *fac* and *mer* isomers.
 Draw these two isomers and label them *fac* and *mer*.

2. Ethylenediamine (en) can act as a bidentate chelating ligand.
  Draw the two structural isomers that can be formed for the complex $[Ir(en)_2Cl_2]^+$.

3. Draw all possible isomers of:
(a) octahedral $[RuCl_2(NH_3)_4]$
(b) square planar $[IrH(CO)(PR_3)_2]$
(c) tetrahedral $[CoCl_3(H_2O)]$
(d) octahedral $[Co(en)(NH_3)_2Cl_2]^+$

4. Give structural formulas, including all isomers, for the following:

(a) Diamminebromochloroplatinum(II)
(b) Diaquadiiododinitropalladium(IV)

5. Give structural formulas, including all isomers, for the following and label the isomers that are cis and trans*, *fac* and *mer***, and enantiomers***:

(a) $[Pt(NH_3)_3Cl_3]^+$ *
(b) $[Co(NH_3)_2(H_2O)_2Cl_2]^+$ ***
(c) $[Co(NH_3)_2(H_2O)_2BrCl]^+$ ***
(d) $[Cr(H_2O)_3BrClI]^+$ ***
(e) $[Pt(en)_2Cl_2]^{2+}$ *   ***
(f) $[Co(NO_2)_3(dien)]$ **

6. The compound $[Co(NH_3)_3(H_2O)Cl_2]^+$ has three isomers where the cis and trans isomers are *mer*, and the other isomer is *fac*. Draw these.

7. The ion $[Co(NH_3)Br(en)_2]^{2+}$ exhibits geometrical isomerism and optical isomerism. Draw these isomers and distinguish between them.

8. How many geometric isomers are possible for the square planar complex $[Pt(NH_3)(OH_2)(Cl)(Br)]$?   Draw them.

9. Determine whether the following complexes have a chiral metal center:
(a) $[Ru(en)_3]^{2+}$
(b) *fac*-$[Rh(en)(H_2O)Cl_3]^+$
(c) *cis*-$[Ir(en)_2Br_2]^+$

10. Use sketches to show that the square planar complexes
(a) $[Pt(NH_3)_2BrI]$ has only two isomers
(b) $[Pt(py)(NH_3)BrCl]$ has three isomers

11. Werner prepared two compounds by reacting a solution of $PtCl_2$ with triethyl phosphine, $P(C_2H_5)_3$. After elemental analysis, the compounds analyzed for Pt 38.8%; Cl 14.1%; C 28.7%; P 12.4% and H 6.02%. Write formulas and structures for the two isomers.

12. What is the relationship between the following two linear complex ions? $[Cl-Ag-SCN]^-$ and $[SCN-Ag-Cl]^-$
The complex ions are:
(a) linkage isomers   (b) coordination isomers (c) geometric isomers (d) optical isomers (e) the same compound

13. Enantiomers are: ____(a) superimposable mirror images with identical chemical formulae and the same chemical reactivities. (b) nonsuperimposable mirror images with identical chemical formulae and the same chemical reactivities. (c) nonsuperimposable mirror images with dissimilar chemical formulae but similar chemical reactivities. (d) superimposable mirror images with identical chemical formulae and similar physical properties

14. What is the relationship between the following two complex ions?

A. $[Co(NH_3)_4(NO_2)Cl]^+$
B. $[Co(NH_3)_4(ONO)Cl]^+$

The complex ions are:
(a) coordination isomers. (b) optical isomers. (c) linkage isomers. (d) geometric isomers. (d) the same compound.

15. How many geometric isomers exist for the tetrahedral complex ion $[CoCl_2Br_2]^{2-}$?

16. Draw the octahedral cobalt ammonates that have their ligands in the following numbered spots in italics:

(a) Co(*1,2,5*-NH$_3$)$_3$(*3,4,6*-Cl)$_3$

(b) Co(*1,2,6*-NH$_3$)$_3$(*3,4,5*-Cl)$_3$

Which of these compounds is *fac* and which is *mer*?

17. In the octahedral cobalt compound $[Co(en)_2(Cl)_2]+$, the isomers that has the ligands in the following numbered places $[Co(3,4-en)(2,5-en)(Cl)_2]+$ should be labeled cis, trans, fac, or mer?

18. Numerically label all the ligands for the enantiomers of $[Co(NH_3)_2(Br)_2(Cl)_2]^-$. Make the two chloride ligands occupy the 1,2 spots for each of these enantiomers.

19. Draw and numerically label all the ligands for the cis, cis, cis and the trans, trans, trans isomers of $[Co(NH_3)_2(Br)_2(Cl)_2]^-$.

Chapter 5.   Crystal Field Theory

Before discussing Crystal Field Theory it is instructive to note the deficiencies of Valence bond theory, and to examine electron counting formalisms.

Valence-Bond Theory versus Crystal Field Theory

Valence bond theory is built on the Lewis electron dot concept that covalent bonding results from sharing of electron pairs between atoms, and then put into a quantum-mechanical approach that was refined by Linus Pauling. The valence bond approach is very useful to organic chemists who use the theory very successfully, but it has been used for years by inorganic chemists---not so successfully.

The main new concept of valence-bond theory is ***the orbital hybridization concept***, which asserts that the wave functions of atomic orbitals of an atom (usually the central atom) can mix together during bond formation. The result is a ***hybrid orbital***. The number of hybrid orbitals formed by this approach equals the number of atomic orbitals that are used to form the hybrid orbitals. Not only can s and p orbitals be utilized, *but d orbitals can be used* as well.

Table 5.1 Hybridization of Central Atom Orbitals.

Hybridization	# of hybrid orbitals	Molecule geometry
sp	2	Linear
sp^2	3	Trigonal planar
sp^3	4	Tetrahedral
dsp^3	5	Trigonal bipyramidal
d^2sp^3	6	**Octahedral**

The formation of hybrid orbitals using valence-bond theory can be used to successfully explain and account for molecular shapes. Unfortunately however, there is <u>no actual evidence</u> for hybridization of orbitals, and the concept of hybridization of atomic orbitals is not a good predictive tool, and this is *especially true* for transition metal complexes.

Valence-bond theory cannot explain magnetic and color properties of transition metal complexes or the spectrochemical series of ligands.

Electron Counting

To be able to use Crystal Field Theory (CFT) successfully, it is essential that one can determine the electronic configuration of the central metal ion in any complex. This requires being able to recognize all the species making up the complex and knowing whether the ligands are neutral or anionic, so that one can determine the oxidation state of the metal ion. This is why we spent time looking at Lewis electron dot structures and ligands in the previous chapters.

In many cases the oxidation state for first row transition metal ions will be either (II) or (III), but in any case you may find it easier to start with the M(II) from which you can easily add or subtract electrons to get the final electronic configuration.

A simple and fun procedure exists for the M(II) case.

First write out all the first row transition metals with their symbols and atomic numbers:

22	23	24	25	26	27	28	29
Ti	V	Cr	Mn	Fe	Co	Ni	Cu

To see the number of electrons in the 3d orbitals, then cross off the first 2, hence:

2	3	4	5	6	7	8	9

So, the electronic configuration of **Ni(II) is d^8** and the electronic configuration of **Mn(II) is d^5**. What is the electronic configuration of Fe(III)? Well, using the above scheme, Fe(II) would be d^6, so by subtracting a further electron to make the ion more positive, the configuration of **Fe(III) will be d^5**.

This simple procedure works fine for first row transition metal ions, but it doesn't for 2nd row elements. (It is interesting to note however that the 3rd row transition elements starting with hafnium through to mercury do have the same last digits as the 2nd row transition elements!)

Electron Counting Formalisms

In the chapter on atomic structure in freshman chemistry we say that in the 4th period of the periodic table the orbitals are filled in the order $[Ar]4s^2 3d^{10}$ for the transition metals. This turns out to be true only for isolated metal atoms that are not in a compound. If a metal ion is surrounded by an electronic field---by surrounding it with ligands---then the 3d-orbitals drop in energy and fill before the 4s orbitals.

Thus, an organometallic chemist naturally considers that the transition metal valence electrons are all d-electrons.

If we ask for the d-electron count on a transition metal such as Nb in the zero oxidation state, we call it d^5, not d^3. For zero-valent metals, we see that the electron count corresponds to the column it occupies in the periodic table. Therefore, Co is in the ninth column and is d^9 (not d^7) and Tc^{3+} is d^4 (seventh column for Tc, and subtract three electrons). Since we can now assign a d-electron count to a metal center, we can determine the electronic contribution of the surrounding ligands in a metal complex and come up with an overall electron count.

Metal Complex Electron Counting Method A:
The Ionic Charge Convention (ICC Method)

The basic premise of this metal-complex electron counting method is that we first remove all of the ligands from the metal with the proper number of electrons for each ligand to bring it to a ***closed valence shell state***, which is usually an octet as we learned in Chapter 2.

For example, if we remove water from a metal complex, H_2O has a completed octet and acts as a neutral molecule. When it was bonded to the metal center it did so through its lone pair and there is no need to change the oxidation state of the metal to balance charge. We therefore call water a neutral two-electron donor, just as we did in Chapter 3.

By contrast however, if we remove a chlorine group from the metal and complete its octet, then we formally have chloride (Cl$^-$). If we bond this chloride anion to the metal, the new lone pair forms our metal-chlorine bond and the <u>chloride ion acts as a two-electron donor</u> ligand. We must do the same thing with a ligand such as a methyl group -CH_3. In the case of a methyl group, the -CH_3 must come away from the metal as a methanide anion (CH_3^-), so that the carbon can maintain an octet.

It is important to notice in these last examples using anionic ligands that to keep electrical charge neutrality we must formally oxidize the metal by one electron by assigning a positive charge to the metal. This reduces the d-electron count of the metal center by one electron.

**Metal Complex Electron Counting Method B:
The Neutral Covalent Convention (<u>NCC</u> Method)**
The basic premise of this metal-complex electron counting method is that when we remove one of the ligands from the metal, rather than take the ligand to a closed shell state, we make it *neutral*. Let's consider water once again. When we remove it from the metal, it is a neutral molecule with one lone pair of electrons. Therefore, as with the ionic charge convention method, water is a neutral two-electron donor.

We diverge from the ionic method when we consider a ligand such as chlorine or methyl. When we remove chlorine from the metal and make the chlorine fragment neutral, we have a neutral chlorine **radical** (Cl·) rather than a chloride ion. Both the metal and the chlorine radical must donate one electron each to form the metal-ligand bond.

Therefore, the <u>chlorine group is a one-electron donor</u>, (as we stated back in Chapter 3. when we first looked at the ways in which to classify ligands) **not** a two-electron donor as it is under the ionic convention. We must do the same thing with a ligand such as the methyl group -CH_3. In the case of a methyl group, the -CH_3 must come away from the metal as a methyl radical (CH_3·), so that the methyl group is electrically neutral.

In the neutral covalent convention, metals retain their full complement of d electrons because we never change the oxidation state from zero when we count electrons; i.e. Co will always count for 9 electrons regardless of the oxidation state and Nb will always count for five.

It is important to notice that this method does not give us any information about the formal oxidation state of the metal. Thus, we must go back and assign that in a separate step, and it is for this reason that ***many chemists prefer the ionic charge convention***. Chemists that are fluent in both methods of electron counting have a greater understanding of the true nature of metal-ligand bonding.

Using the ICC Method to Calculate Oxidation State

In a metal complex the oxidation state of the metal is calculated by determining the difference in the number of valence electrons in the zero oxidation state of the metal with the number of electrons present after the loss or removal of the ligands in their closed Lewis electron dot structures.

For example we can calculate the oxidation state of the metals for: $[MnO_4]^-$, $[PtF_6]$, $[Cr(NH_3)_6]^{3+}$, $[Cr(CO)_5]^{2-}$.

For $[MnO_4]^-$: Mn^o has 7 e-'s,   7
      overall charge on the complex is -1,   +1
      remove 8- charges for four O^{2-} - 8
      electrons left on the Mn   0

The difference between the number of valence electrons on the metal (7) minus the number after removal of ligands (0): $(7) - (0) = +7$ is the oxidation state of Mn. Thus Mn(VII)

For $[PtF_6]$: Pt^o has 10 e-'s,   10
      overall charge on the complex is 0 ,   0
      remove 6- charges for six F-   - 6
      electrons left on the Pt   +4

The difference between the number of valence electrons on the metal (10) minus the number after removal of ligands (0): $(10) - (4) = +6$ is the oxidation state of Pt. Thus Pt(VI).

For $[Cr(NH_3)_6]^{3+}$: Cr^o has 6 e-'s,   6

         overall charge on the complex is +3 ,   -3

         remove 0 charges for CO   <u>0</u>

         electrons left on the Cr   +3

The difference between the number of valence electrons on the metal (6) minus the number after removal of ligands (3): $(6) - (3) = +3$ is the oxidation state of Cr. Thus Cr(III).

For $[Cr(CO)_5]^{2-}$: Cr^o has 6 e-'s,   6

         overall charge on the complex is -2 ,   +2

         remove 0 charges for CO   <u>0</u>

         electrons left on the Cr   +8

The difference between the number of valence electrons on the metal (6) minus the number after removal of ligands (8): $(6) - (8) = -2$ is the oxidation state of Cr. Thus Cr^{-2} .

Crystal Field Theory

There are two things that set the study of the electronic structures of transition metal compounds apart from the body of valence theory.

One is the presence of partly filled d subshells. This leads to experimental observations not possible in most organic chemistry such as: paramagnetism, visible absorption spectra, and apparent irregular variations in thermodynamic and structural properties.

The second is that there is a crude but effective approximation, called ***crystal field theory*** that provides a powerful yet simple method of understanding and correlating all of those properties that arise primarily from the presence of the partly filled d subshells.

The crystal field theory provides a way of determining how the energies of the metal ion orbitals will be affected by the set of surrounding atoms or ligands by simple electrostatic considerations.

Crystal field theory is a *model* and not a realistic description of the forces actually at work, and is completely superseded by molecular orbital theory. With that being said, its simplicity and convenience have earned it a cherished place in the coordination chemist's "toolbox."

The electronic properties of transition metal complexes are discussed in terms of the "orbital splittings," which the crystal field theory enables us to work out relatively easily. Our attention will be confined entirely to the d-block elements, and will be focused primarily on those of the 3d series. This is where the crystal field theory works best. The splittings of f orbitals are generally so small that they are not chemically important.

In crystal field theory, it is assumed that the metal ions are **simple point charges** (cations with a positive charge) and the ligands are also simple point charges (anions with a negative charge). When applied to alkali metal ions containing a symmetrical sphere of charge, calculations of energies are generally quite successful. The approach taken uses classical potential energy equations that take into account the attractive and repulsive interactions between charged particles (that is, Coulomb's Law interactions).

Electrostatic Potential is proportional to $q_1 \times q_2/r^2$, where q_1 and q_2 are the charges of the interacting ions and r is the distance separating them. This leads to the correct prediction that large cations of low charge, such as K^+ and Na^+, should form few coordination compounds.

For transition metal cations that contain varying numbers of d electrons in orbitals that are NOT spherically symmetric, however, the situation is quite different. The shape and occupation of these d-orbitals then becomes important in an accurate description of the bond energy and properties of the transition metal compound.

To be able to understand and use CFT then, it is absolutely essential to have a clear picture of the shapes of the d-orbitals.

One of the main differences between the three *p* orbitals and the five *d* orbitals is that the set of *p* orbitals has three identical orbitals oriented along the *x, y,* and *z* axes whereas the set of five *d* orbitals has four identical orbitals ($d_{xy}, d_{yz}, d_{xz},$ and $d_{x^2-y^2}$) and one (the d_{z^2}) that *looks* like it is special; that is, it *appears* to be rather different from the other four. This distinction between the d_z^2 orbital and the other four needs to be addressed in some detail, and follows after the next paragraph.

The following scheme illustrates the differences in the d-orbitals. Make sure that you understand and remember which d-orbitals lie along which axes of the Cartesian coordinate system, and which d-orbitals lie between the axes. This is the most important distinction among the five d-orbitals, not their shapes!

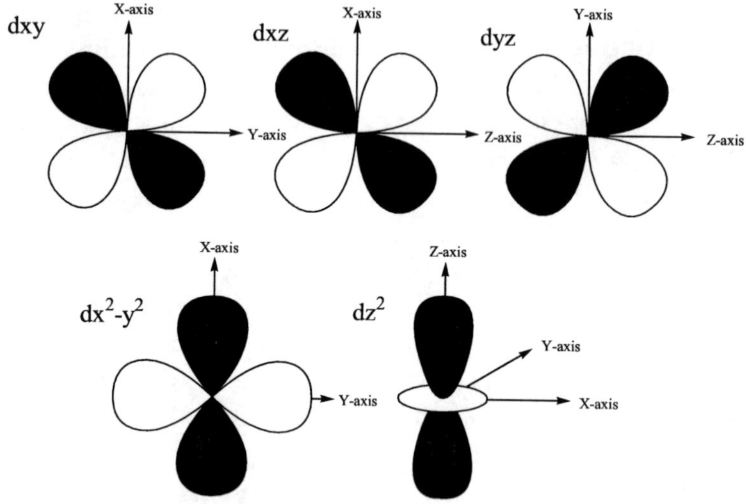

The Schrödinger mathematics, which yields the probability of finding electrons in various orbitals, can be carried out a number of different ways yielding a variety of solutions. *The set of five 3d orbitals depicted above is just one possible set of orbitals.*

Another solution of the Schrodinger mathematics in fact yields *six* dependent orbitals. These six dependent *d* orbitals are the d_{xy}, d_{yz}, d_{xz}, $d_{x^2-y^2}$ orbitals shown above, plus two more similar to $d_{x^2-y^2}$, labeled $d_{z^2-y^2}$ and $d_{z^2-x^2}$.

The five 3*d* orbitals that we normally use are generated from the six *dependent* orbitals by taking a linear combination of (that is by adding) the $d_{z^2-y^2}$ and $d_{z^2-x^2}$ orbitals to generate the d_z^2 orbital. When these two orbitals are added together, the resulting d_z^2 orbital has twice as great an electron probability along the *z*-axis as it does along the other two axes. Keep in mind that the d_z^2 orbital is not as special as it looks; it is merely a linear combination of two dependent orbitals that look exactly like the other four orbitals in the *d* subshell.

The scheme below shows the five d-orbitals situated inside the Cartesian coordinate system (drawn with a hypothetical cube) at the origin with the x, y, and z axes labeled. Again, as discussed on the previous page, the $d_{x^2-y^2}$ and the d_z^2 orbitals lie along the axes, and the d_{xy}, d_{yz}, d_{xz} orbitals do not.

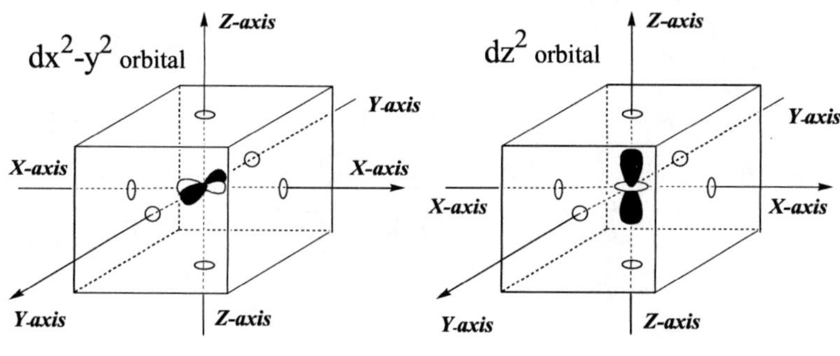

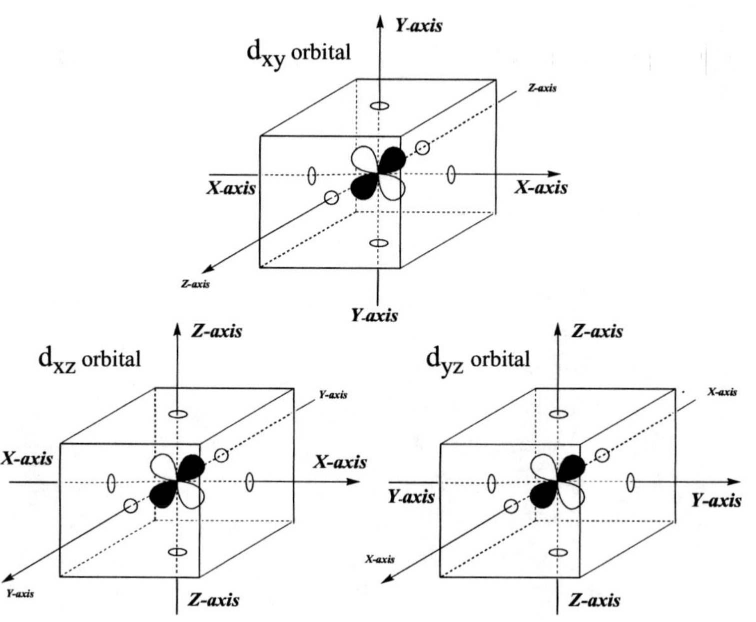

Octahedral Geometry

Now, let us play a mind game and build a metal complex up from a free, gaseous metal ion with no ligands around it. The five d-orbitals in a gaseous metal ion are degenerate and have the same energy; this is illustrated in the following energy diagram.

If a spherical field of negative charges (a so-called crystal field) is placed around the metal ion, then the energy of the d-electrons in the d-orbitals is increased because of electrostatic repulsion (the d-electrons are negative charged just as the negative charges placed around the metal ion so they repel each other), but the five d-orbitals are still degenerate.

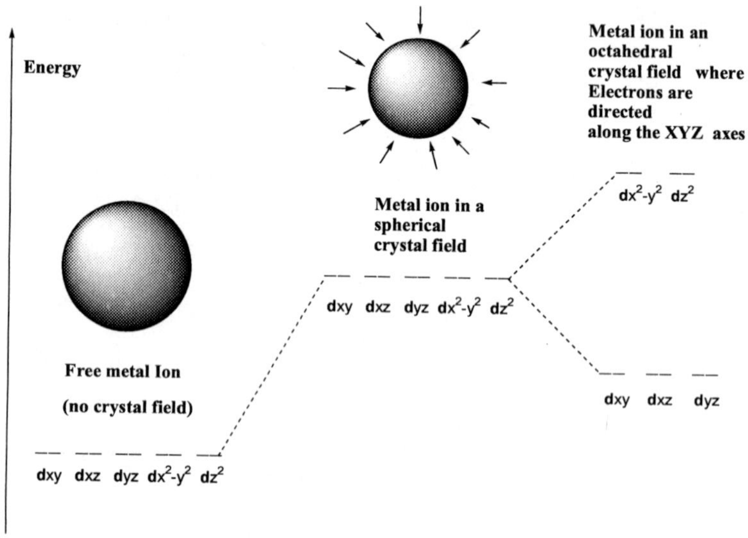

This is shown in the diagram above by an increased energy level, as compared to the free metal ion, for the five d-orbitals, but they are all still degenerate.

Next, let us replace the spherical field of negative charge with a directed field. That is, let us move in six point charges, or ligands if you will, directly along the three axes of the Cartesian coordinate system, the X, Y, and Z-axes. This is the same geometry as an octahedral shaped metal complex, for instance $TiCl_6^{3-}$, or $[Co(NH_3)_6]^{3+}$.

In this octahedral case, things change dramatically. The five d-orbitals are no longer degenerate, and electrons in those d-orbitals that have lobes extending along the X, Y, and Z axes (dx^2-y^2 and dz^2) will experience greater repulsion due to these "incoming" ligands as compared to the electrons in the d-orbitals that have lobes that lie between the X, Y, and Z axes (*dxy, dyz* and *dxz*).

This is called *crystal field splitting*.

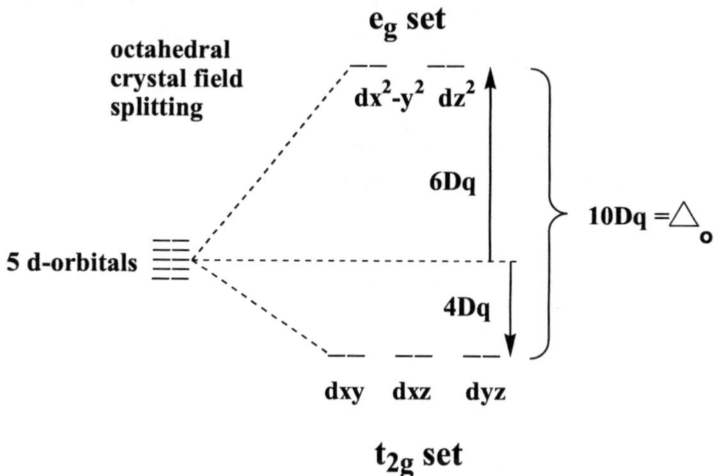

The result of these differences for the d-orbitals is an energy difference between the dz^2 and dx^2-y^2 orbitals compared to the *dxy, dyz* and *dxz* orbitals called **10Dq** by definition, or Δ_o, which can vary in energy depending on metal and ligands, but is typically in the range of 100-300 kj/mol.

This splitting energy is typified by an increase in energy for the two orbitals lying along the axes, called the **e_g set**, and a decrease in energy for the three orbitals lying in between the axes, called the **t_2g set**.

The e_g set is *increased* in energy by 6 Dq from the "barycenter" and the t_2g set is *decreased* by 4 Dq.

As mentioned above, the value of 10 Dq varies for different metal complexes.

Let's look at some metal complexes that exhibit these differences.

Consider the simple example of $TiCl_6^{3-}$ in which six chloride ions are in an octahedral arrangement around the Ti^{3+} cation, which is d^1.

There is only one d-electron to be allocated to one of the five d- orbitals.

If it were to occupy the dz^2 or dx^2-y^2 orbital, both of which point directly towards the chloride ligands, it would be strongly repelled.

The geometry of the dz^2 or dx^2-y^2 orbitals and their nodes would require the electron to stay near the negatively charged ligands---causing even *more repulsion* than a spherically distributed electron would experience.

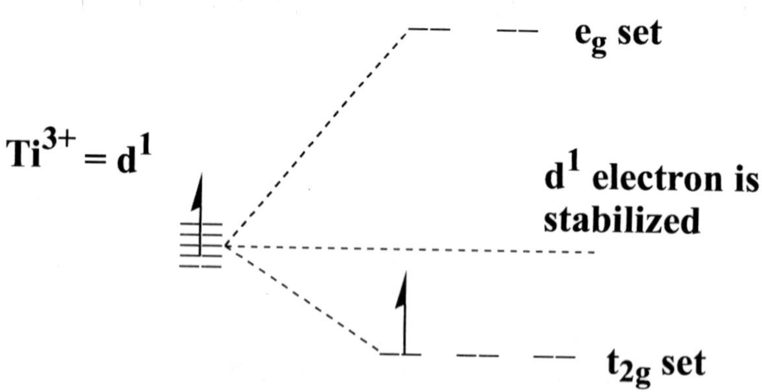

On the other hand, if the electron were to occupy the dxy, dyz or dxz orbital, it would spend less time near the ligands than would a spherically distributed electron and would be repelled less.

Thus, the d^1 electron is stabilized by 4Dq by CFT.

In the case described above for $TiCl_6^{3-}$ where there is only one d-electron, the electron goes into one of the t_{2g} orbitals, and this has the effect of stabilizing the electron by -4 Dq.

This stabilization by 4 Dq is called the **crystal field stabilization energy (CFSE)**.

For a metal complex that has two d-electrons (d^2), then the CFSE would be twice as much, or -8Dq, (See HUND's RULE).

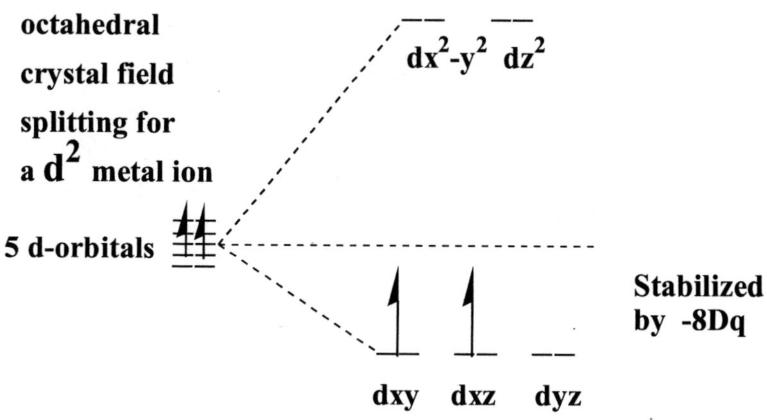

Electrons that are admitted into the t_{2g} set are therefore stabilizing for octahedral metal complexes.

An octahedral metal complex with three d-electrons (d^3) would then have a CFSE that would be -12 Dq.

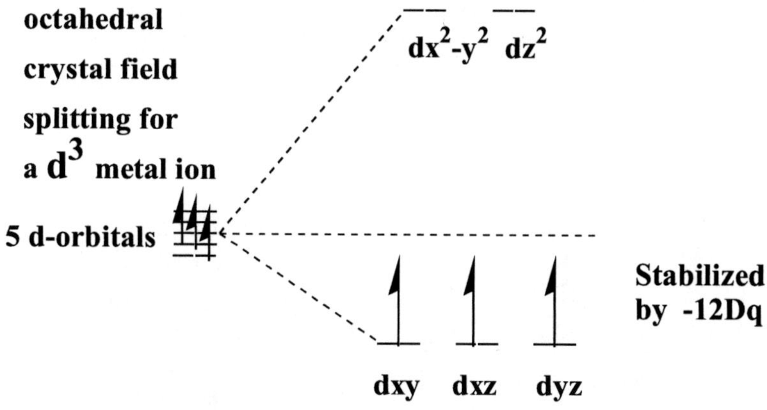

In the cases where there are four d-electrons (d^4) there are two possibilities that arise. *In the case where 10Dq is less than the pairing energy* for two electrons in the same orbital, the fourth electron will occupy one of the higher energy e_g orbitals, and *in the case where 10Dq is greater than the pairing energy*, the fourth electron will pair up in one of the t_{2g} orbitals.

These are called the **weak field case** and the **strong field case** respectively. These cases are shown below with a relative illustration of the orbital energy splitting.

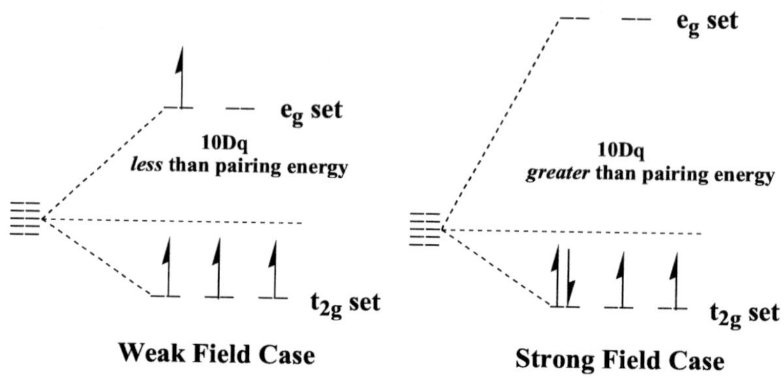

The option for weak field or strong field is only available with d^4 through d^7 metal complexes. (See Table 5.2)

You should satisfy yourself that this statement is true by making a set of orbital energy diagrams like those above and fill in electrons for the situations where there are 8 to 10 d-electrons.

For the cases shown above there are different numbers of paired and unpaired d-electrons. For the weak field case, there are four unpaired electrons, whereas in the strong field case there are only two unpaired electrons. The electron configurations can be written in this manner:

Weak field $d^4 =$ $t_{2g}^{3} e_g^{1}$ is also called **high spin**
Strong field $d^4 =$ $t_{2g}^{4} e_g^{0}$ is also called **low spin**

Therefore, the weak field case is also called *high spin*, and the strong field case is also called *low spin* because of the lower number of unpaired electrons.

The two possible cases where an octahedral metal complex can have a d^5 electron configuration are shown below:

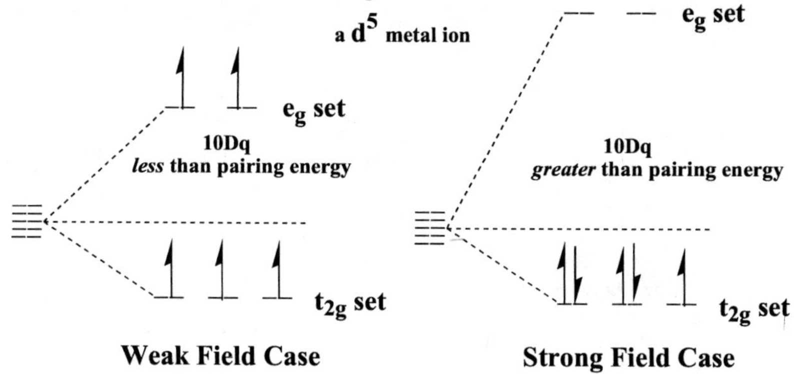

The d^6 and d^7 electron configurations are shown below:

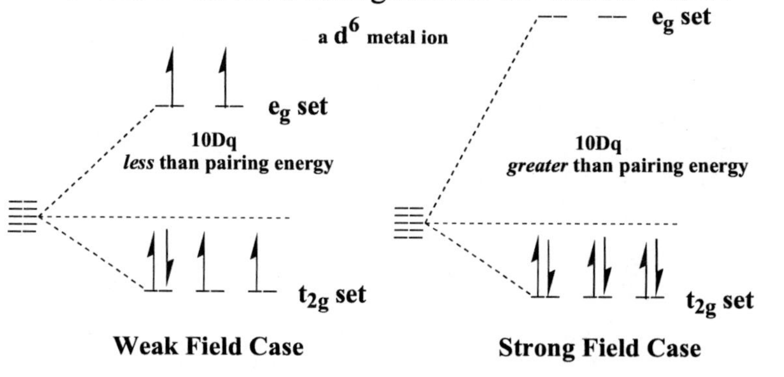

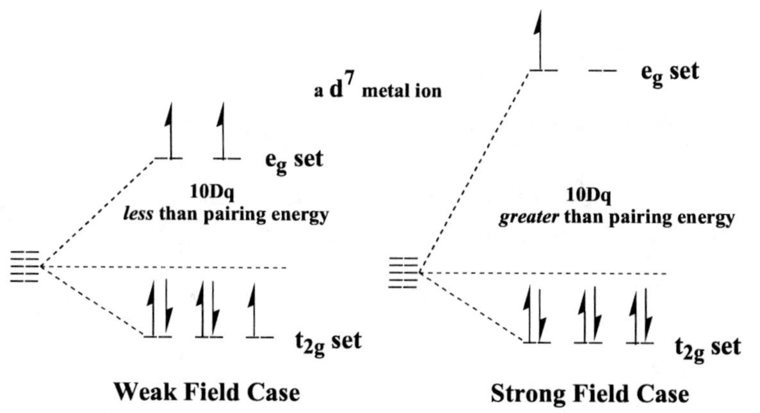

- The d^6 strong field case has the maximum CFSE.

The d^8 and d^9 configurations show **NO DIFFERENCE** in their electron configurations in the weak and strong field cases; the electron configuration for d^8 is $t_{2g}^6 e_g^2$, and for d^9 it is $t_{2g}^6 e_g^3$.

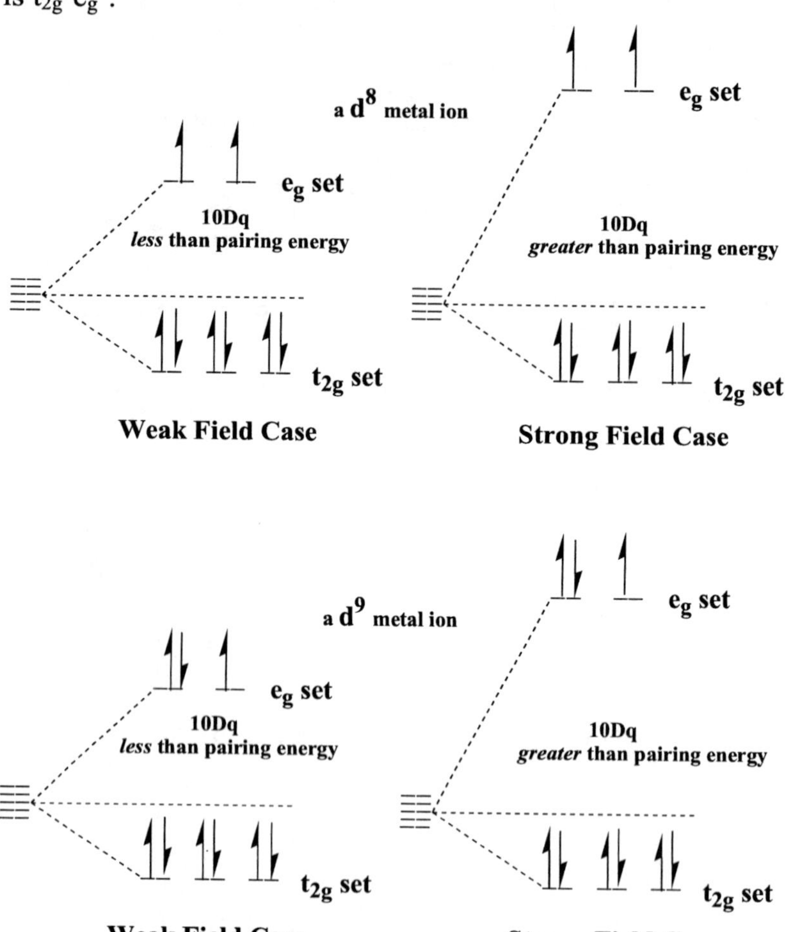

Of course, it goes without saying that the d^{10} configuration is the same no matter what the field strength or ligand.

The table below shows the crystal field effects for weak and strong octahedral fields. The P stands for the *pairing energy*.

Table 5.2 Crystal Field Stabilization Energies for the Octahedral Geometry

WEAK FIELD			STRONG FIELD		
dn	Config	CFSE	dn	Config	CFSE
d^1	$t_{2g}^1 e_g^0$	-4Dq	d^1	$t_{2g}^1 e_g^0$	-4Dq
d^2	$t_{2g}^2 e_g^0$	-8Dq	d^2	$t_{2g}^2 e_g^0$	-8Dq
d^3	$t_{2g}^3 e_g^0$	-12Dq	d^3	$t_{2g}^3 e_g^0$	-12Dq
d^4	$t_{2g}^3 e_g^1$	**-6Dq**	d^4	$t_{2g}^4 e_g^0$	**-16Dq + P**
d^5	$t_{2g}^3 e_g^2$	**0Dq**	d^5	$t_{2g}^5 e_g^0$	**-20Dq + 2P**
d^6	$t_{2g}^4 e_g^2$	**-4Dq + P**	d^6	$t_{2g}^6 e_g^0$	**-24Dq + 3P**
d^7	$t_{2g}^5 e_g^2$	**-8Dq + 2P**	d^7	$t_{2g}^6 e_g^1$	**-18Dq + 3P**
d^8	$t_{2g}^6 e_g^2$	-12Dq + 3P	d^8	$t_{2g}^6 e_g^2$	-12Dq + 3P
d^9	$t_{2g}^6 e_g^3$	-6Dq + 4P	d^9	$t_{2g}^6 e_g^3$	-6Dq + 4P
d^{10}	$t_{2g}^6 e_g^4$	0Dq + 5P	d^{10}	$t_{2g}^6 e_g^4$	0Dq + 5P

Tetrahedral Geometry

The CFT approach can be easily extended to other geometries and the next most important case is **tetrahedral**. The Δ_{td} splitting of the d-orbitals is much smaller than Δ_o on the order of: $\Delta_{td} = 4/9\Delta_o$ since there are only four ligands around a tetrahedral complex compared to six on octahedral complexes, which reduces overall repulsion, and since the ligands are not as efficiently directed at the d-orbitals.

The scheme below depicts the Cartesian coordinate system with a hypothetical cube drawn around it, and then next a tetrahedral metal complex at the origin of the coordinate system, where the ligands are found at the corners of the hypothetical cube. Since the ligands are at the corners of the cube, no d-orbitals are directed towards them.

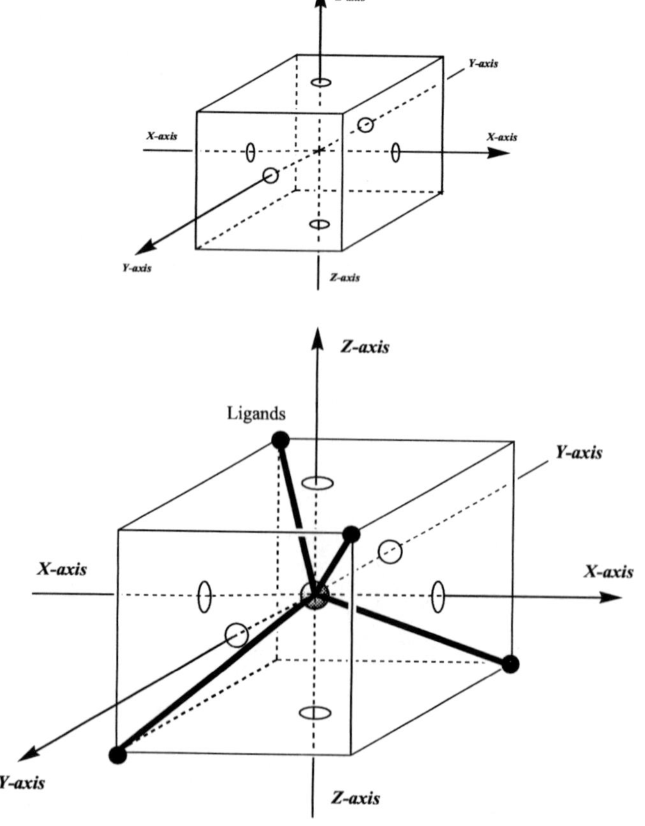

This can be illustrated by drawing a couple of representative d-orbitals into this scheme. For illustrative purposes and brevity, only the dz^2 and the dxz orbitals are shown, representative of the e set and the t set of orbitals.

As can be discerned from the complicated scheme below, the e set of orbitals (dz^2 and the dx^2-y^2) have a lesser interaction with the ligands (1/2 face of cube away) than does the t set of orbitals (dxy, dyz, dxz) (1/2 of edge away).

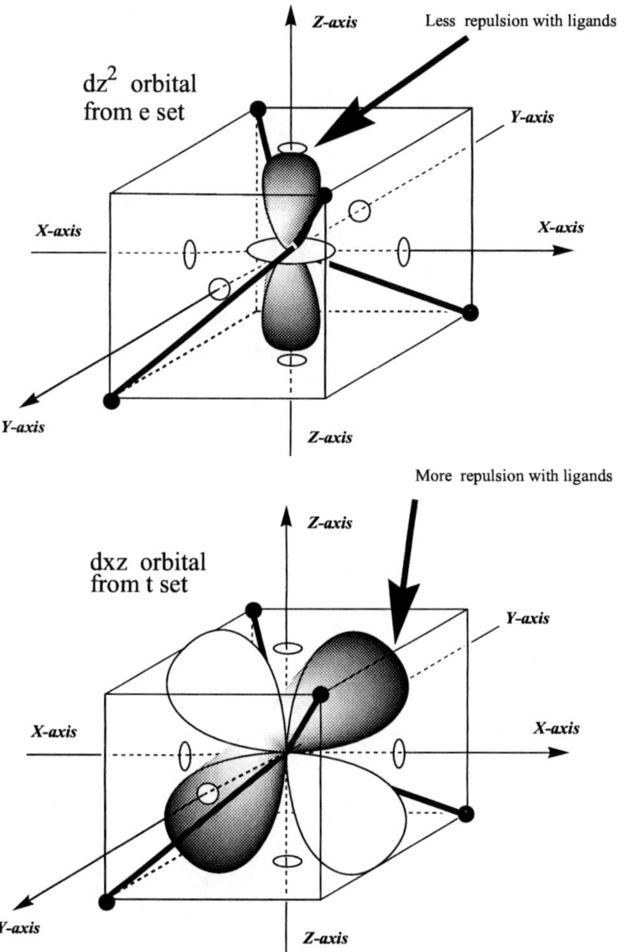

Since this is the reverse of the octahedral case, the d-orbital splitting diagram is reversed as seen on the next page. Overall, the d-orbital splitting is not nearly as strong as in the octahedral case!

Since Δ_{td} is so small, *tetrahedral metal complexes are always considered to be weak field-high spin cases.*

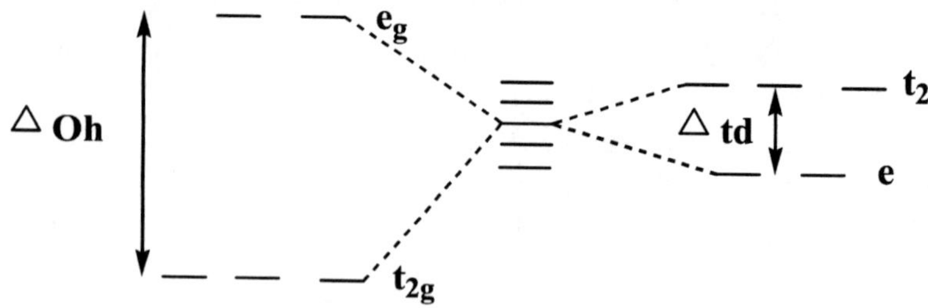

This can be illustrated with the tetrahedral complex ion $[CoCl_4]^{2-}$. The ion contains Co^{2+}. You should be able to determine that for yourself at this point. The Co(II) ion has a d^7 electron configuration. The splitting of the d-orbitals will be small since this is a tetrahedral complex, giving the weak field/ high spin case shown in the scheme below.

The compound $[CoCl_4]^{2-}$ exhibits *paramagnetism*, with three unpaired electrons.

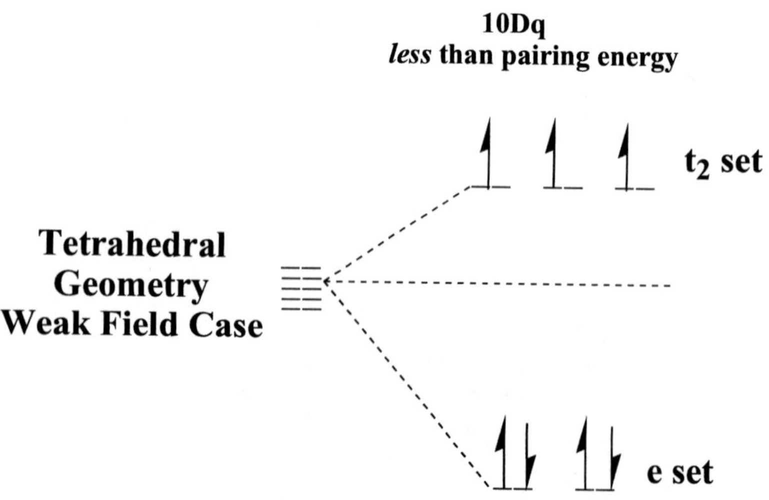

The **square planar case** can be considered as an extension of the octahedral field, where we remove the two ligands from the Z-axis.

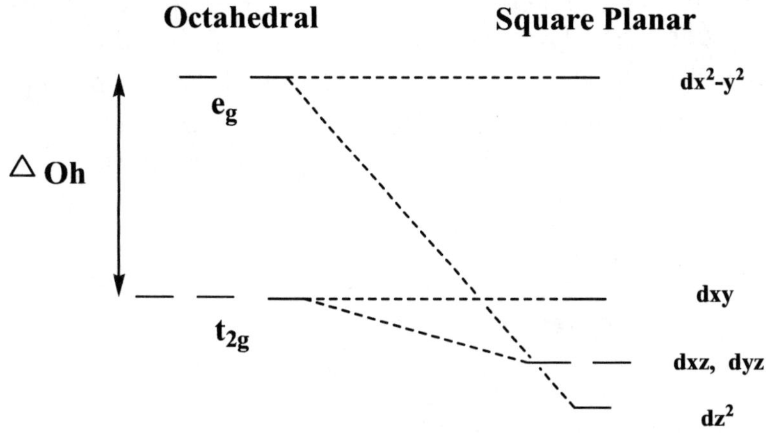

Consequently, repulsion of an electron in the dz^2 orbital will no longer be equivalent to that experienced by an electron in the dx^2-y^2 orbital, and the end result is shown above. In this text, the only square planar complexes will be for d^8 complexes, i.e. nearly all four coordinate complexes exhibit tetrahedral geometry except for d^8 metal complexes, which may be **tetrahedral or square planar**.

An example of a square planar complex is $[Ni(CN)_4]^{2-}$. The Ni(II) is d^8, so the splitting diagram is:

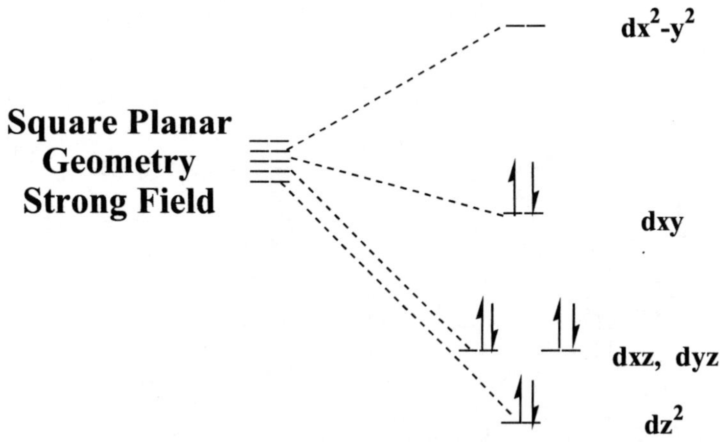

In this case, there are no unpaired electrons; the compound is considered to be *diamagnetic*.

Jahn-Teller Effect

The ***Jahn-Teller theorem*** states that any species with an electronically degenerate ground state will distort to remove the degeneracy. Compounds exhibit the Jahn-Teller effect by displaying a distortion of their coordination geometry as the degeneracy of the ground state is "broken".

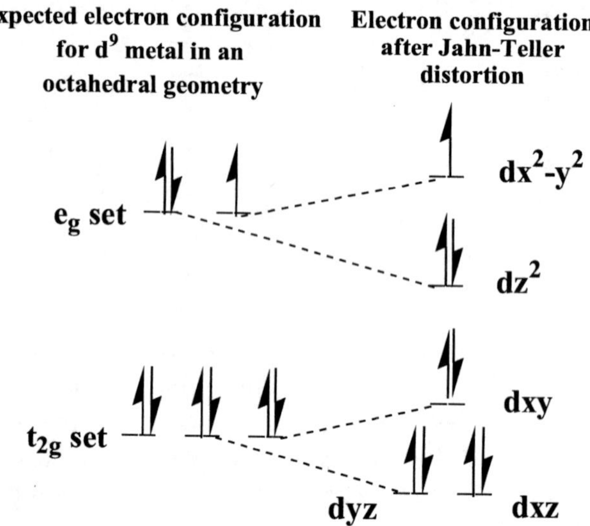

For transition metal complexes the Jahn-Teller effect is observed particularly for d^9 metal complexes in an octahedral field, and one of the best examples is the $[Cu(H_2O)_6]^{2+}$ ion. The bonds along the z-axis are longer, "*lifting the degeneracy*" of the dz^2 and the dx^2-y_2 orbitals.

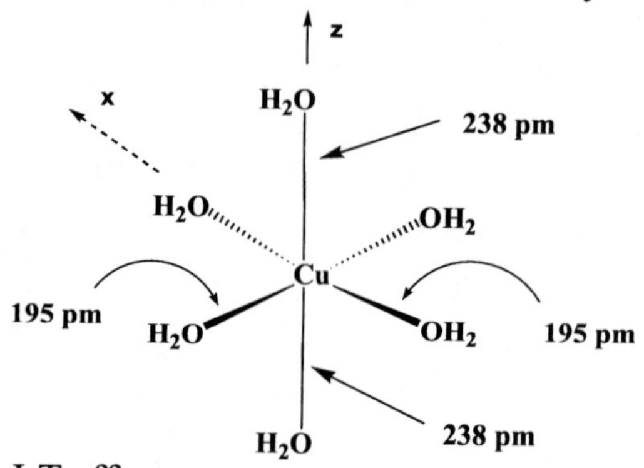

Why is the J-T effect also seen in some high spin d^4 and low spin d^7 cases?

Trends in Crystal Field Theory

A trend in the value of Δ_o can be determined for metal ions which is independent of the ligands. This trend is: Mn(II) < Ni(II) < Co(II) < Fe(III) < Cr(III) < Co(III) < Ru(III) < Mo(III) < Rh(III) < Ir(III) < Pt(IV). The trend follows roughly the oxidation state of the metals (+3 has a higher field than +2 for the same metal and ligand), then it increases down a group.

Generally, Δ_o increases with increasing oxidation state of the metals but, this is an empirical generalization and **crystal field theory just cannot explain the magnitudes of Δ_o values**. Generally, complexes of the 2nd and 3rd row of the periodic table almost always form strong field complexes, so Crystal Field Theory is best for 1st row transition metal complexes.

Magnetism

Compounds are considered to be diamagnetic when they have no unpaired electrons, but paramagnetic when they have one or more unpaired electrons. Since electrons have negative charge and are moving around an atom this constitutes an electric field, and electric currents set up magnetic fields.

An atom will exhibit a magnetic field when it is paramagnetic, but in a diamagnetic compound the paired electrons in an orbital cancel each other out so that no resultant magnetic field can be measured.

Paramagnetic compounds align these unpaired electrons in an applied external magnetic field, and this is how paramagnetism is measured. *Paramagnetic compounds are attracted to an applied external magnetic field.*

Diamagnetic compounds oppose the applied magnetic field (***Lenz's Law***) and are thus repelled by the applied external magnetic field.

The result is that paramagnetic compounds are attracted to an applied external magnetic field and become apparently heavier when weighed on a magnetic susceptibility balance, and diamagnetic compounds are repelled by an applied external magnetic field and become apparently lighter when weighed on a magnetic susceptibility balance.

Magnetic moments can then be calculated for each compound measured on a magnetic susceptibility balance. To a first approximation the magnetic moments can be calculated for a metal complex by using the number of unpaired electrons (n).

The following table shows some calculated and observed values for the magnetic moments of some metal complexes with their number of unpaired electrons.

Table 5.3 Calculated and Observed Values of Magnetic Moments

Complex	n	$\mu_{Calculated}$	$\mu_{Observed}$
$[Cu(H_2O)_6]^{2+}$	1	1.73	1.75
$[Ni(H_2O)_6]^{2+}$	2	2.83	2.83
$[Co(H_2O)_6]^{2+}$	3	3.87	4.85
$[Mn(H_2O)_6]^{3+}$	4	4.90	4.93
$[Fe(H_2O)_6]^{3+}$	5	5.92	5.40

One method of determining the number of unpaired electrons is by looking at the magnetic properties of the compounds. A simple technique to determine a magnetic moment (the GOUY METHOD using a Gouy balance) involves weighing the sample in the presence and absence of a strong magnetic field. By careful calibration using a known standard, such as $HgCo(SCN)_4$ the number of unpaired electrons can be determined.

Color

Colored compounds absorb visible light. The color perceived is the sum of the light that isn't absorbed by the complex. (This means that we see the color of a metal complex as the complimentary color of the absorbed species of light!) The amount of absorbed light versus wavelength is an absorption spectrum for a complex.

The color of a transition metal complex depends on the metal, its oxidation state, and its ligands. The pale blue transition metal species $[Cu(H_2O)_4]^{2+}$ can be converted into the dark blue $[Cu(NH_3)_4]^{2+}$ by adding $NH_3(aq)$. A partially filled set of d orbitals is usually required for a complex to be colored. *So, d^0 and d^{10} metal ions are usually colorless.* Exceptions to this are MnO_4^-, which is Mn(VII), and CrO_4^{2-}, which is Cr(VI). These colors are from ***charge-transfer***.

For example, the absorption spectrum for $[Ti(H_2O)_6]^{3+}$ has a λ_{max} (maximum absorption) at 510 nm, which is in the green-yellow part of the visible spectrum, so, the complex transmits all light except green-yellow. Therefore, we see the complex as the complimentary color of this yellow-green absorption, which is purple. See the color wheel on the front cover of the book. (The absorption peaks are often very broad so the colors expected from comparison with the color wheel may not match exactly.)

For some various transition metal species:

~Color absorbed	~Wavelength absorbed	~Color observed
Violet	410nm	yellow
Blue-green	460nm	orange
Green	500nm	red
Red	660nm	blue

Since $E=h\nu$ and $E=hc/\lambda$, then the shorter the wavelength (λ) absorbed, the larger is Δ_o.

Kinetics

The rate of ligand exchange can be fast for a metal complex (***kinetically labile***) or it can be very slow (***kinetically inert***). The rate of ligand exchange can depend on many factors such as the exchange mechanism, type of ligand, solvent, temperature, etc. An important factor is the crystal field stabilization energy of a metal complex.

If the metal is highly stabilized, such as Co^{3+}, which has t_{2g}^6 electron configuration (low spin) and 24Dq CFSE, or Cr^{3+} which has t_{2g}^3 electron configuration (low spin) and 12Dq CFSE, then it is kinetically inert. Complexes with other electron configurations such as d^4 and d^9 are very kinetically labile. Why?

Inert $= d^3$, d^6, d^8 (low spin)   Labile $= d^4$, d^9

Limitations of the Crystal Field Theory

The Crystal Field Theory is a metal-based theory, and does not delve into the effects of various ligands on the crystal field strength. Generally, Δ_o increases with increasing oxidation state of the metals, however, this is an empirical generalization and crystal field theory just cannot explain the magnitudes of Δ_o values. Real progress in understanding can only be made by considering the bonding between ligands and metals.

There are other major problems with the crystal field theory. Ligands are not point charges, but crystal field theory treats ligands as point charges or dipoles. Even charged ligands such as the chloride ion have a polarized charge distribution when bonded to transition metal ions, so they can't be considered to be point charges either.

Crystal field theory does not take into account the overlapping of ligand orbitals with metal-based orbitals in a more covalent fashion, such as in the bonding of neutral molecules such as carbon monoxide with transition metals.

Another problem is the lack of distinction between strong field ligands and weak field ligands. Why does a ligand such as chloride and water form weak field complexes, yet ligands such as the organophosphines and carbon monoxide form strong field complexes?

The crystal field theory has no explanation for these observations, and as a matter of fact, it fails completely. *How can the anionic Cl⁻ ion function as a weak field ligand when a neutral covalent molecule such as carbon monoxide acts as a strong field ligand?*

Clearly the crystal field theory is incomplete and either needs to be modified or discarded. Since it is so useful, it has been modified and has evolved instead into *Ligand Field Theory.*

Ligand Field Theory

Van Vleck adopted Ligand Field Theory in 1935 as a modification of Crystal Field Theory in order to account for more covalency in bonding of ligands to metal, which is lacking in the purely electrostatic Crystal Field Theory. *It has been established that the ability of ligands to cause a large splitting of the energy between the orbitals is essentially independent of the metal ion.*

The ***Spectrochemical Series*** is a list of ligands ranked in order of their ability to cause large orbital separations. A shortened list includes:

$$I- < Br- < \underline{S}CN- \sim Cl- < F- < OH- \sim \underline{O}NO- < C_2O_4^{2-} <$$
$$H_2O < \underline{N}CS- < EDTA^{4-} < NH_3 \sim pyr \sim en < bipy < phen$$
$$<< PR_3 < CN- \approx NO \approx CO$$

Take, for example, two different metal complexes of Fe^{2+}- d^6, which are: $[Fe(H_2O)_6]^{2+}$ and $[Fe(CN)_6]^{4-}$.

<table>
<tr><td align="center">Weak Field Case
High Spin</td><td align="center">Strong Field Case
Low Spin</td></tr>
</table>

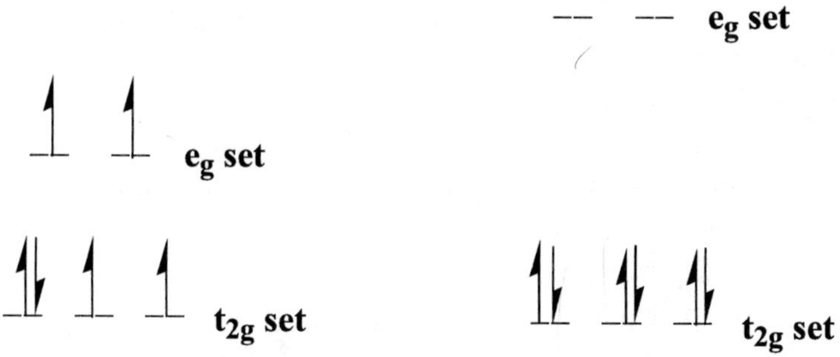

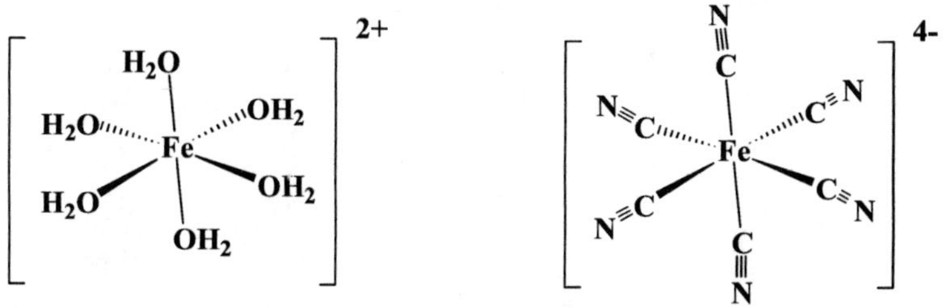

The diagram on the left represents the case for the aqua ion (small Δ) and on the right that of the hexacyano ion (large Δ).

What this means is that if one uses a technique, such as magnetic susceptibility, that can detect the presence of unpaired electrons in each compound, then one will find that $[Fe(H_2O)_6]^{2+}$ has four unpaired electrons, while in the latter compound $[Fe(CN)_6]^{4-}$ there will be no unpaired electrons.

This accounts for the terms **high spin** (most unpaired electrons and most paramagnetic) and **low spin** (least unpaired electrons).

The terms weak field and strong field give an indication of the splitting abilities of the ligand.

Water always gives rise to small splittings of the energy levels of the d orbitals for first row transition metal ions and hence is referred to as a weak field ligand. Conversely, CN^- is a strong field ligand, since it causes large splittings of the energy levels of the d-orbitals.

How do you know if a ligand is strong or weak field without the use of the spectrochemical series in hand? **The general rule of thumb is that CO, CN⁻, phosphines and NO⁺ are all strong field ligands.** The halides are weak field ligands, and the nitrogen-based ligands are probably weak field but can be strong field ligands in certain cases. We will discuss this further in Chapter 6, and explain the spectrochemical series using molecular orbital theory.

As has been mentioned by other inorganic chemists, the valence-bond theory picture and the ligand field theory picture can both be considered to be specializations of the molecular orbital method.

Some aspects of the material in the next chapter on MO theory are within the boundaries of Ligand Field Theory. We have decided to combine the material and present the quantum mechanical aspects of Ligand Field Theory and Molecular Orbital Theory at one time in the same chapter.

Terms and Definitions for Chapter 5:

Orbital Hybridization Concept
Hybrid orbitals

Electronic Configuration

Crystal Field Theory

Paramagnetism

Visible Absorption Spectra
Charge Transfer
Simple Point Charges
Electrostatic Potential
Cartesian Coordinate System

Crystal Field Splitting
10dq
e_g Set
t_{2g} Set
Crystal Field Stabilization Energy (CFSE)
Weak Field Case
Strong Field Case
Hund's Rule
High Spin
Low Spin

Kinetically Labile
Kinetically Inert

Diamagnetic
Paramagnetic

Ligand Field Theory
Spectrochemical Series

Jahn-Teller Theorem

Chapter 5 Problems and Exercises

1. What is the color of an aqueous solution containing $Zn(NH_3)_4^{2+}$? (a) blue (b) yellow (c) red (d) colorless

2. Which pair has a d^{10} electron configuration?
(a) Cu^{2+} and Ni^{2+} (b) Co and Ni
(c) Zn and Cu^{2+} (d) Cu and Zn (e) Cu+ and Zn^{2+}

3. What are the electron configurations of Fe^{2+} and Co^{3+}, respectively?
(a) both d^7 (b) d^7 and d^8 (c) both d^6
(d) d^6 and d^8 (e) both d^5

4. Which of the following complexes contains a d^3 metal ion? Give the number of d-electrons for each metal.
    (a) $K_2[Ni(CN)_4]$
    (b) $[Ni(NH_3)_6]Cl_2$
    (c) $K_4[Fe(CN)_6]$
    (d) $[Co(OH_2)_6](ClO_4)_2$
    (e) $[Cr(OH_2)_5Cl]Cl_2$

5. Predict the number of unpaired electrons in $[Cr(CN)_6]^{4-}$ and $[Cr(OH_2)_6]^{2+}$, respectively. Draw the d-orbital splitting diagrams.

6. Predict the number of unpaired electrons in the tetrahedral complex ion $[MnCl_4]^{2-}$. Draw the d-orbital splitting diagram.

7. Rank the following complex ions from the lowest d-orbital splitting energy to the highest.
(a) $[Co(OH_2)_6]^{3+}$ (b) $[Co(ox)_3]^{3-}$ (c) $[CoCl_6]^{3-}$
(d) $[Co(NH_3)_6]^{3+}$ (e) $[Co(CN)_6]^{3-}$

8. Rank the following complex ions from the lowest d-orbital splitting energy to the highest.
(a) $[Fe(NH_3)_6]^{2+}$ (b) $[Fe(CN)_6]^{4-}$ (c) $[Fe(OH_2)_6]^{2+}$
(d) $[FeCl_6]^{4-}$ (e) $[FeCl_4]^{2-}$

9. For which one of the following would it not be possible to distinguish between high-spin and low-spin complexes in octahedral geometry?
(a) Ni(II) (b) Co(III) (c) Fe(II) (d) Co(II) (e) Cr(II)

10. The complex ion $[Cr(OH_2)_6]^{2+}$ has four unpaired electrons. Which statements are true?
(a) the complex is high-spin.
(b) the complex is low-spin.
(c) the complex is diamagnetic.
(d) Δ_o is very large.
(e) the water ligands are difficult to remove.

11. Predict the total number of d-electrons in a complex having one unpaired electron in a strong octahedral field and three unpaired electrons in a weak octahedral field.

12. An octahedral complex is high-spin when the value of Δ_o is _________________ than the energy required to _________________ the electrons. (hint-Hund's rule)

13. Determine which of these metal complexes are high spin.
$[CoF_6]^{3-}$ $\mu_{Obs} = 5.3$
$[Fe(CN)_6]^{3-}$ $\mu_{Obs} = 2.3$
$[Co(NO_2)_6]^{4-}$ $\mu_{Obs} = 1.8$

14. Which complex ion should absorb visible light of the shortest wavelength, $[Cr(H_2O)_6]^{3+}$, or $[CrCl_6]^{3-}$?

15. For Co^{3+}, the pairing energy is 252 kJ/mol and the d-orbital splitting energies produced by F- ligands is 155 kJ/mol, and the d-orbital splitting energies produced by NH_3 ligands is 276 kJ/mol. Sketch the d-orbital splitting diagrams for the two complexes $[CoF_6]^{3-}$ and $[Co(NH_3)_6]^{3+}$, and place electrons in the correct d-orbitals. Which are high spin or low spin complexes?

16. On the basis of the given colors for the following complexes, list the ligands NH_3, H_2O, Cl-, and NCS- in order of increasing field strength, and then compare your list with the spectrochemical series.

$[Co(NH_3)_6]^{3+}$	orange-yellow
$[Co(NH_3)_5Cl]^{2+}$	purple
$[Co(NH_3)_5(NCS)]^{2+}$	orange
$[Co(NH_3)_5(H_2O)]^{3+}$	red

17. Rank the following complex ions in order of increasing Δ_o and the energy of visible light absorbed: $[Cr(NH_3)_6]^{3+}$, $[Cr(H_2O)_6]^{3+}$, $[Cr(Br)_6]^{3-}$.

18. Rank the following complex ions in order of decreasing Δ_o and the energy of visible light absorbed: $[Cr(en)_3]^{3+}$, $[Cr(CN)_6]^{3-}$, $[Cr(Cl)_6]^{3-}$.

19. The complex $[Cr(H_2O)_6]^{3+}$ is violet. A second CrL_6 complex is green. The ligand is probably… (a) CN-   (b) Cl-

20. Draw a d-orbital diagram for the Mn^{3+} ion (high spin). Would you suspect that Jahn-Teller distortion could occur?

21. Draw a d-orbital diagram for the Ni^{3+} ion (low spin). Would you suspect that Jahn-Teller distortion could occur?

22. Explain why coordination compounds containing Cu^{2+} are colored but coordination compounds containing Cu^+ are not.

23. Complexes with the electron configurations d^4 and d^9 are usually very kinetically *labile*. Why?

24. Complexes with the electron configurations d^3 and d^6 are usually very kinetically *inert*. Why?

Practice Quiz

1. What is the oxidation state (use roman numerals please) of the metal and the number of d electrons in:

$[CoCl_6]^{3-}$ = __________ and it has ________ d electrons
$[Cr(CN)_6]^{4-}$ = __________ and it has ________ d electrons
$[CuCl_4]^{2-}$ = __________ and it has ________ d electrons
$[Cr(OH_2)_6]^{2+}$ = __________ and it has ________ d electrons

2. Draw the five d-orbitals with respect to their alignment with the x,y, and z-axes.
Label the e_g and the t_{2g} set.

3. Predict the number of unpaired "d" electrons in $[W(CN)_6]^{4-}$ and $[W(OH_2)_6]^{2+}$, respectively. Draw the d-orbital splitting diagram for these octahedral complexes.

______4. Predict the number of unpaired electrons in the tetrahedral complex ion $[CuCl_4]^{2-}$.
(A) 1 (B) 2 (C) 0 (D) 3 (E) 4

______5. Which of the following complex ions has the smallest d-orbital splitting energy (Δ_o) ?
(A) $[Rh(OH_2)_6]^{3+}$ (B) $[Rh(ox)_3]^{3-}$ (C) $[RhCl_6]^{3-}$
(D) $[Rh(NH_3)_6]^{3+}$ (E) $[Rh(CN)_6]^{3-}$

______6. Which of the following complex ions has the highest d-orbital splitting energy (Δ_o) ?
(A) $[Os(NH_3)_6]^{2+}$ (B) $[Os(CN)_6]^{4-}$ (C) $[Os(OH_2)_6]^{2+}$
(D) $[OsCl_6]^{4-}$

7. For what numbers of d-electrons would it not be possible to distinguish between high-spin and low-spin complexes in octahedral geometry?

8. The complex ion $[Mo(OH_2)_6]^{2+}$ has four unpaired electrons.

   This means that (circle all that are correct)....
   (A) the complex is high-spin.
   (B) the complex is low-spin.
   (C) the complex is diamagnetic.
   (D) the complex is paramagnetic
   (E) Δ_o is larger than the pairing energy
   (F) Δ_o is smaller than the pairing energy.

_______9. Predict the total number of d-electrons in a complex having two unpaired electrons in a strong octahedral field and four unpaired electrons in a weak octahedral field.
(A) 7 (B) 5 (C) 6 (D) 8 (E) 4

____________10. An octahedral complex is high-spin, weak field when the value of Δ_o is larger than the energy required to pair up the electrons. (TRUE or FALSE)

11. How many unpaired electrons would $[Tc(CN)_6]^{3-}$ and $[TcCl_6]^{4-}$ have?
Give electronic configurations for each ion in terms of the t_{2g} and e_g sets. Carefully rationalize your answers by drawing the d-orbital splitting diagrams for each one and tell if it is high field (large Δo) or weak field (small Δo).

BONUS question: $Na_2[Pt(CN)_4] \cdot 3H_2O$ is diamagnetic. Speculate on the basic geometry of the complex anion, $[Pt(CN)_4]^{2-}$, by using a d-orbital crystal field splitting diagram.   Briefly explain your answer. Is it tetrahedral or square planar?

Chapter 6.   Molecular Orbital Theory

The molecular orbital approach spreads electrons out over many atoms (like resonance) and thus has the effect of giving a more stable bonding situation for molecules, and even transition metal complexes.

To be worthwhile, the molecular orbital approach has to explain bonding for weak and strong field ligands, and also not lose the usefulness of crystal field theory. Some of the strongest aspects of the molecular orbital theory are that it can explain the bonding situation for metal carbonyls and strong field ligands and explain the overall spectrochemical series.

In MO theory, we will start with looking at combinations of atomic orbitals on the ligands and the metal center and see what kind of orbital overlaps are allowed. This is only an introductory level college course textbook, so some readers will be unfamiliar with a lot of molecular orbital theory. That is okay, this is presented and based on material about atomic orbitals (s, p, d) that the reader should have been exposed to in a freshman chemistry course, and then we'll go from there.

In Ligand Field Theory, along with basic Molecular Orbital Theory, we start our systematic discussion of bonding by constructing ***linear combinations of atomic orbitals*** (LCAO's) based on atomic orbitals from each of the bonding atoms.

The best way to present this material is to first look at the simplest systems, such as the simple diatomic molecules. This allows us to systematically build more complex sets of LCAO's, and help us to distinguish between sigma and pi interactions between these linear combinations of atomic orbitals.

The difference between core and valence electrons can then also be distinguished in this manner, and we can then direct our attention on the outermost LCAO's, which are often called the "frontier orbitals".

σ (sigma) Bonding

Electron density lies between the nuclei in molecular compounds for any sigma bond. A sigma bond may be formed by linear combinations of **s-s, p-p, s-p, s-d, p-d, d-d,** or hybridized orbital overlaps, and shortly we will work through examples of the first two of these (s-s, and p-p).

The bonding molecular orbitals produced by the linear combination of atomic orbitals approach are lower in energy than the original atomic orbitals.

The antibonding MO's are higher in energy than the original atomic orbitals.

Bond Order

Bond order is the measure of the stability (or reactivity) of a chemical covalent bond, and electrons in bonding orbitals stabilize the bond, whereas electrons in antibonding orbitals destabilize the bond.

The higher the bond order, the more stable the bond; and the bond order can be calculated by the following equation:

$$BO = \frac{\left(e^- \, in \, bonding \, orbitals\right) - \left(e^- \, in \, antibonding \, orbitals\right)}{2}$$

Simple Inorganic Molecules (diatomic molecules)

For a simple diatomic molecule such as H_2, the molecular orbital description is not all that complicated.

For H_2, the covalent bond is formed by simple overlap of s-orbitals to form a **sigma** (σ) bond. The sigma bond essentially allows the electron pair to travel over a larger area, and thus it has less energy than the simple atomic s-orbitals it is derived from. The scheme below shows the formation of a sigma bond (constructive overlap), and the concomitant antibonding orbital that also forms (destructive overlap).

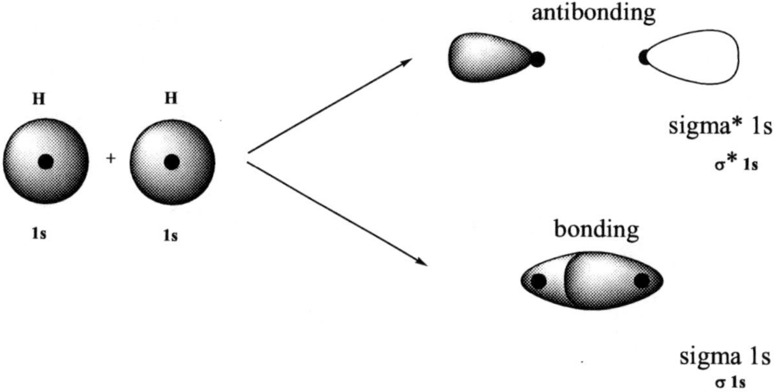

The formation of He$_2$ is not observed in nature and our theory shows that the antibonding orbital would be filled, and therefore there would be no net bonding interaction.

The molecular orbital energy diagrams for H$_2$, and He$_2$ are shown in the following scheme.

Other diatomic inorganic molecules that are commonly found in the environment, or can be readily synthesized are N$_2$, O$_2$, and F$_2$. We drew the Lewis electron dot structures of these molecules earlier on, and we should remember that N$_2$ has a triple bond when the Lewis structure is drawn. What would molecular orbital theory tell us? Each nitrogen atom has seven total electrons, but two are used to fill the 1s core, and therefore nitrogen (Group V) has five valence electrons in 2s and 2p orbitals. We discuss this next.

The three p orbitals, when all are shown in a three-dimensional fashion around the central atom, take up much of the space, as can be seen in the scheme below. The three p orbitals, by convention, lie along the x, y, and z axes of the Cartesian coordinate system.

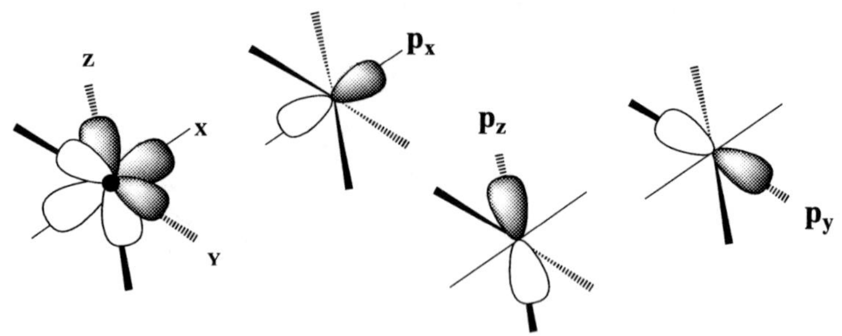

The direct overlap of orbitals between the bonding atoms in molecular orbital theory results in a σ-bond (sigma) regardless of the type orbitals that are used to construct the molecular orbital.

Thus, a σ-bond can be formed by the constructive overlap of two in-phase p orbitals on bonding atoms in a diatomic molecule.

This kind of sigma overlap also results in destructive interference by out-of-phase orbitals to produce a σ * antibonding molecular orbital. These molecular orbitals are shown in the following scheme:

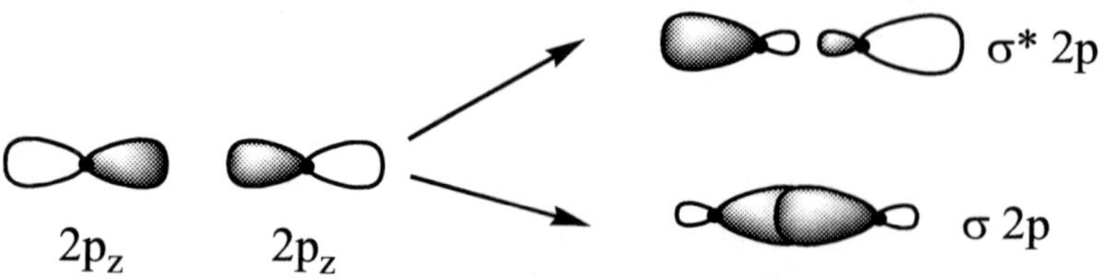

The remaining two p orbitals from one atom in a diatomic molecule cannot directly overlap with the other remaining p orbitals on the second atom to form a sigma bond, but they can overlap in a quite different fashion.

π (pi) Bonding

This side-to-side overlap is known as a pi (π) molecular orbital. The two possible p-π molecular orbitals for a diatomic molecule are degenerate in energy, i.e., they have the same energy. One of the two degenerate p-π molecular orbitals is shown in the following scheme, as well as the resulting antibonding p-π* molecular orbital.

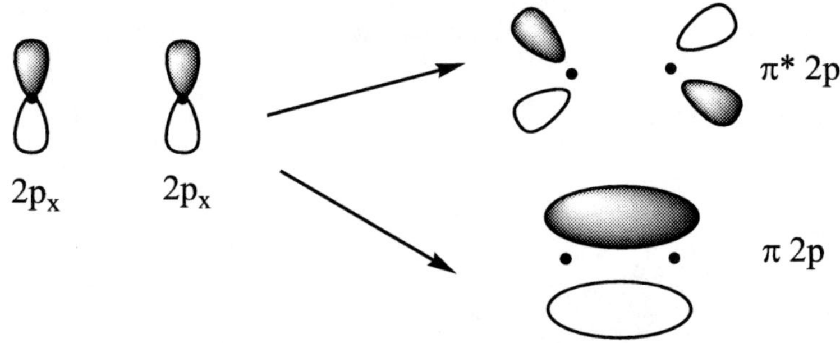

The resulting antibonding p-π* molecular orbitals are consequently of higher energy than the bonding p-π molecular orbitals.

The molecular orbital energy diagram for the p-orbital interactions to form molecular orbitals can now be shown in the following scheme.

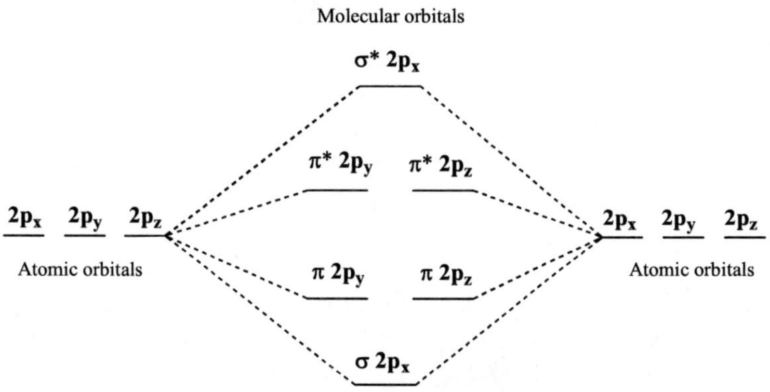

If we now consider a real molecule (lets take the case of F_2) then we must add in the electrons into the molecular orbitals.

Let us now consider all of the electrons for F_2, the core electrons as well as the valence electrons.

The molecular orbital energy diagram shown below illustrates all of the points we have discussed so far.

The bond order is 1. This can be calculated using the bond order equation shown earlier. The molecule has 10 electrons in bonding orbitals and 8 electrons in antibonding orbitals.

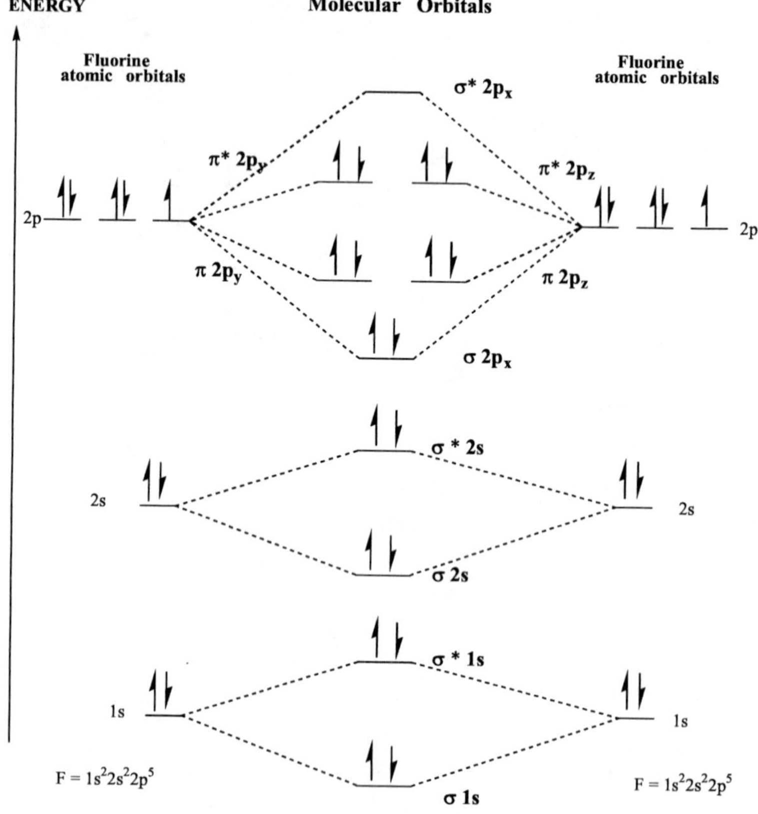

The electrons in LCAO's produced by the 1s orbitals are core electrons. The electrons in the LCAO's produced by the 2p orbitals are in *"frontier orbitals"* which are the *highest occupied molecular orbitals (HOMO)*, and the *lowest unoccupied molecular orbitals (LUMO)* are unfilled.

Molecular oxygen (O_2) has some interesting properties. The Lewis structure shows the presence of a double bond. Draw it.   ::O=O::

Even though the Lewis structure shows the presence of a double bond, and no unpaired electrons, the physical data contradicts this drawing. When placed in a magnetic field, liquid oxygen is attracted to the poles of a magnet. **This shows that O_2 is paramagnetic, and has unpaired electrons!!**

Molecular orbital theory on the other hand predicts that it should have unpaired electrons!!

Thus for molecular O_2:

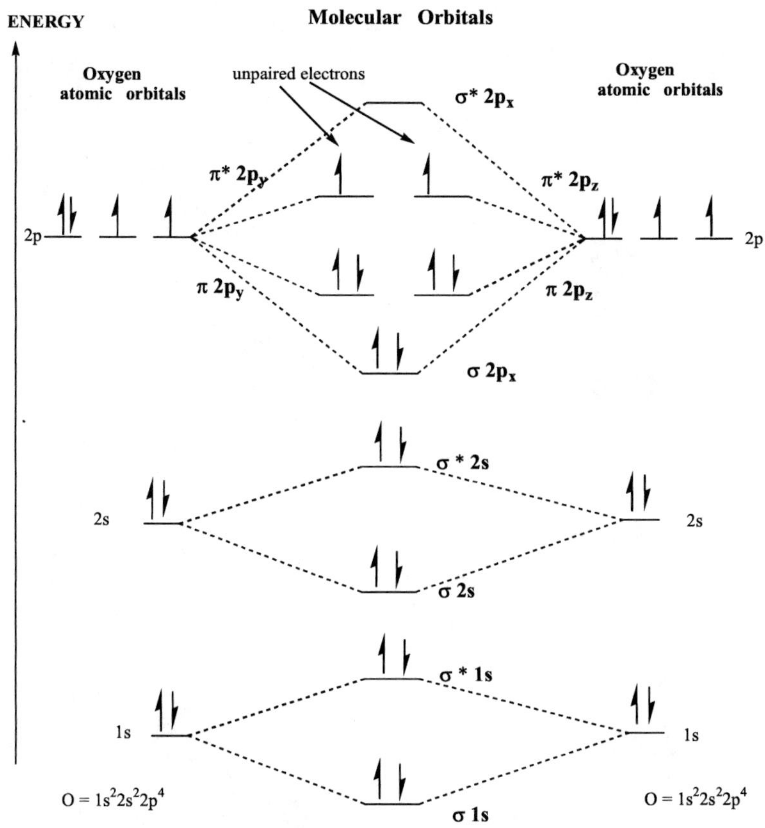

What is the bond order for molecular O_2 using MO theory?

Molecular Orbitals for d-block elements

      The transition metal center has s, p, and d orbitals that are available to provide orbital overlap with ligand-based orbitals. The ligands themselves may have "hybrid orbitals" for the lone pairs of electrons on the ligating atoms, such as sp^3, sp^2, or sp hybrid orbitals, and these can be used to overlap the atomic orbitals on the metal. Orbitals are drawn with different shadings on the different lobes of p and d orbitals, these are called angular overlap shadings, and may be alternatively designated with a + or − sign.

      Orbitals can only overlap and "merge" together if they have the same overlap symbol (said to be "in-phase"), otherwise they are considered to be "out of phase". This is analogous to constructive and destructive interference of waves. So, let us look at that in phase/out of phase situation for different orbitals. As shown in the diagram below there is net overlap for a d orbital with an s orbital, as is the case between d and p orbitals if they are correctly aligned.

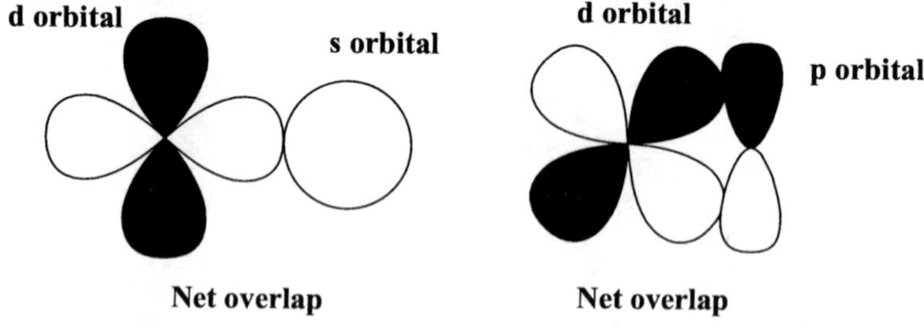

      On the other hand, improper alignment of orbitals will lead to situation where the orbitals cannot merge, and there is no net overlap. This is seen in the scheme below:

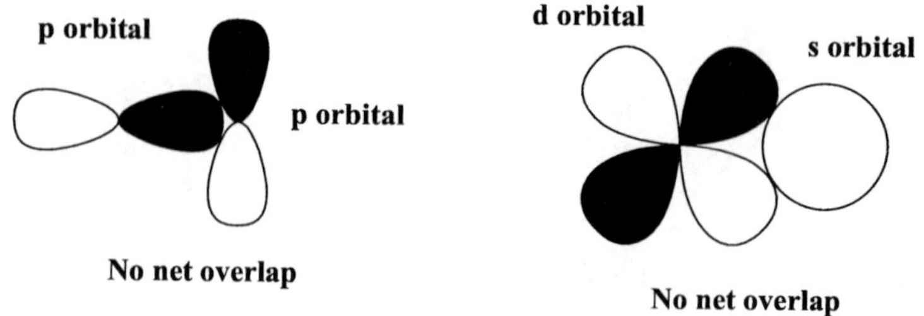

Now let us imagine a more complicated situation. Let us imagine a hypothetical octahedral metal complex with six ligands bound to the metal. MO theory says that we should be able to draw six molecular orbitals that would describe the six bonds. However, unlike valence bond theory, the bonds are not all of the same energy, and the bonds may be spread out over many atoms. Our bonding combination of orbitals is based on symmetry considerations. Thus:

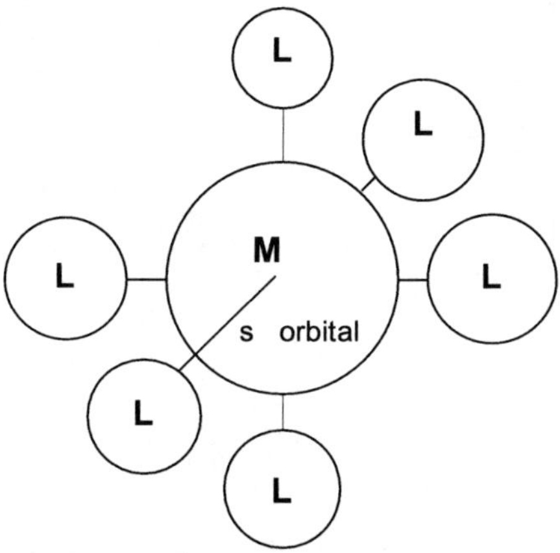

Here the metal s orbital (4s) can merge with six ligand lone pair orbitals (hybrid orbitals) to form a molecular orbital (a_{1g}) bonding combination. This is the lowest energy molecular orbital for bonding between ligands and the metal.

Next, we can construct a set of three molecular orbitals using three p orbitals on the metal, and the ligand orbitals:

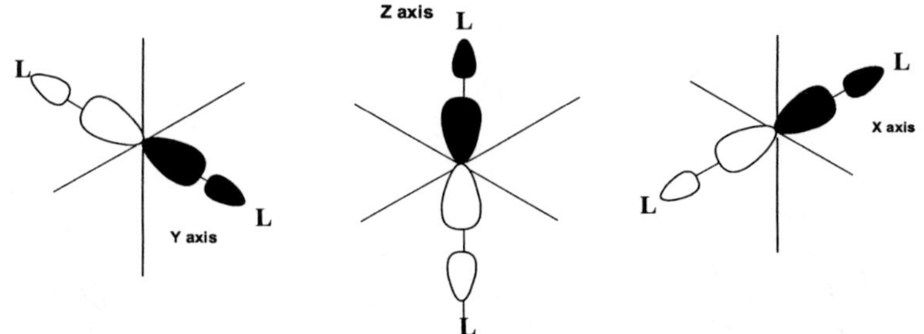

These three molecular orbitals are given the designation t_{1u}.

Lastly, the five d-orbitals form a set of two bonding molecular orbitals (e_g **set** with the dz^2 and the $dx^2\text{-}y^2$), and a set of three non-bonding orbitals (t_{2g} **set** with the **dxy, dxz, and the dyz orbitals**).

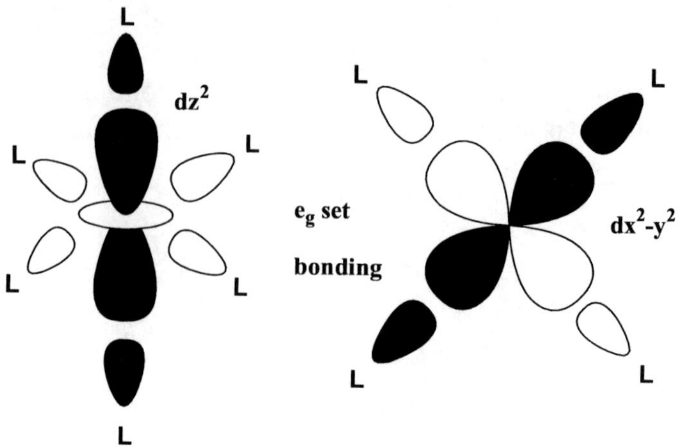

As you can see, there are six bonding molecular orbitals that can be formed from the metal orbitals; there are two bonding molecular orbitals and three non-bonding molecular orbitals.

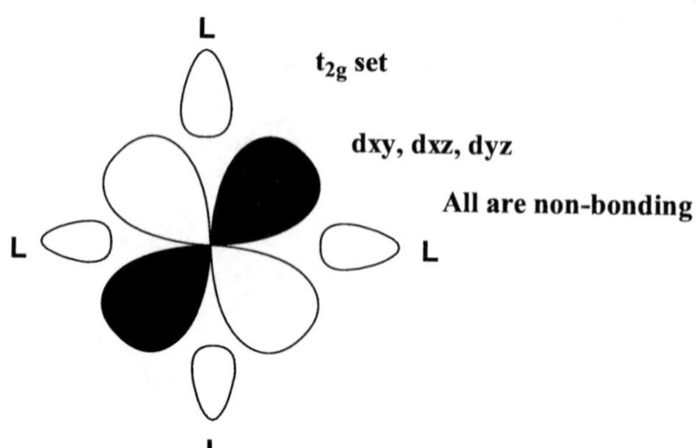

In MO theory for every bonding molecular orbital that you can be constructed, there is a corresponding antibonding orbital of higher energy. The molecular orbitals from lowest energy to highest energy would be bonding, non-bonding, and then antibonding.

σ-base ligands (σ-donor ligands) **[H₂O]**

The following is an energy level diagram for a ML_6 complex showing the sigma-bonding contributions.

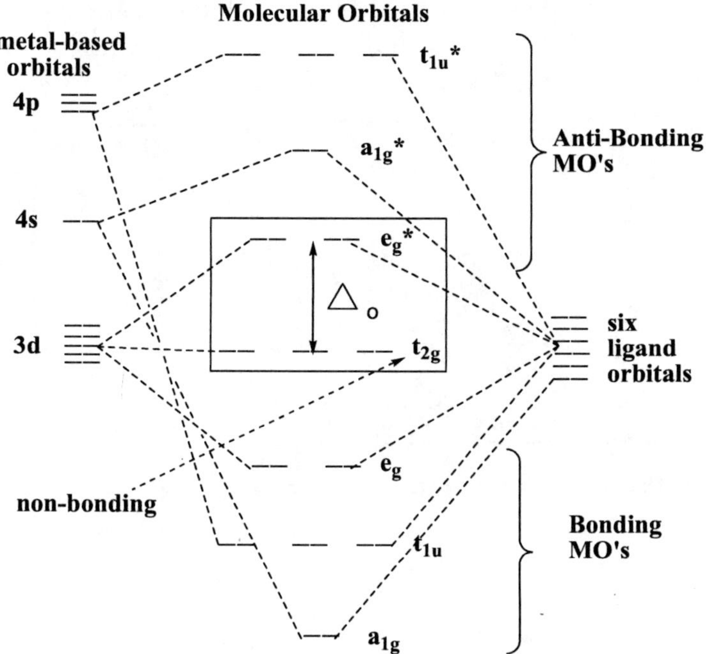

From the MO diagram above, one can see that even though a complicated MO diagram has evolved, the essential crystal field component remains intact, and Δ_o can be seen. The insert showing the frontier orbitals is shown below:

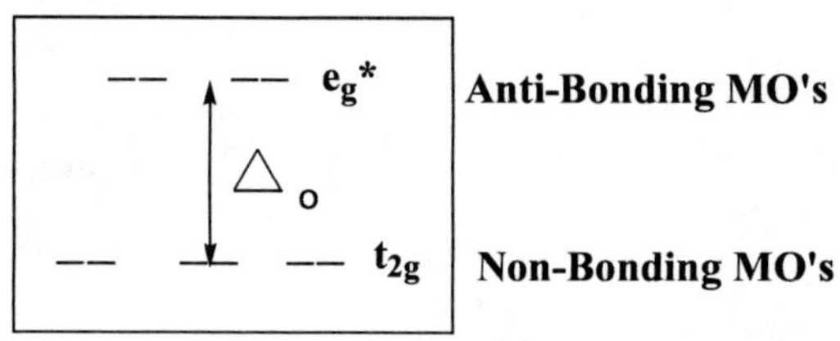

This MO model can be investigated even further for different types of ligands and for different complex geometries, but the main point has been established. MO theory can be developed and yet maintain the useful aspects of crystal field theory. (see problem 5, page 174)

π-base ligands (π-donor ligands/σ-donor) [Cl]⁻

The lone pairs on a ligand such as the chloride ion essentially act as π-donors and contribute electron density onto the t_{2g} orbitals based on the metal. These t_{2g} orbitals become bonding orbitals along with the corresponding formation of the antibonding t_{2g}^* orbitals at higher energy.

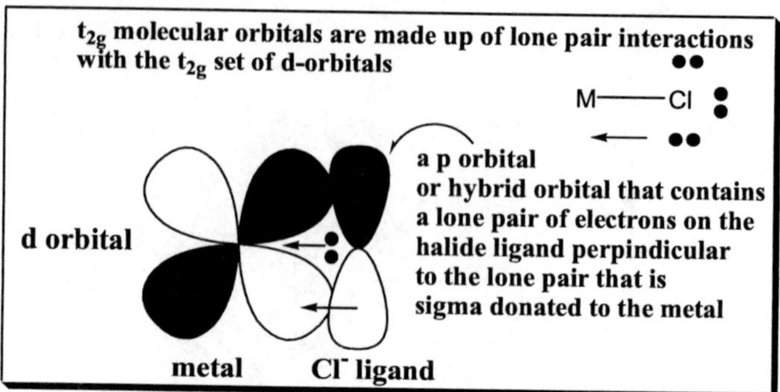

The effect of π-donor ligands such as chloride ion is to decrease Δo. The following is an energy level diagram for a ML_6 complex showing the sigma-bonding contributions, and the π-bonding contributions. (see problem 6, page 174)

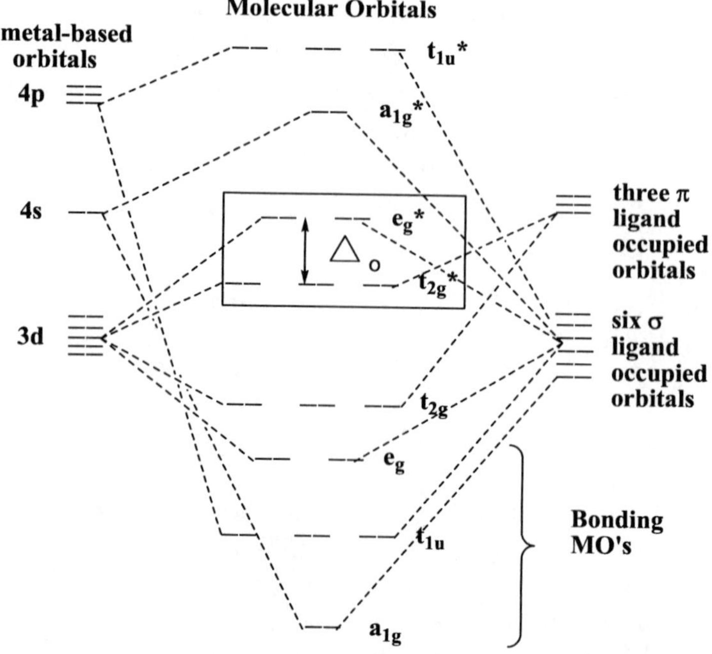

π-acid ligands (π- acceptor ligands/σ-donor) [CO]

The following is an energy level diagram for a ML_6

Molecular Orbitals

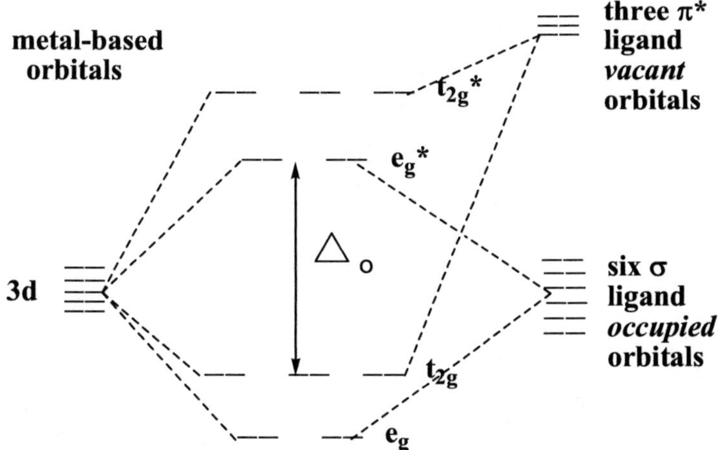

complex showing the sigma-bonding contributions, and importantly, the π-bonding contributions for a π - acceptor ligand. The empty π * orbitals on a ligand such as carbon monoxide act essentially as π-acceptors and siphon off electron density from the t_{2g} orbitals based on the metal.

The effect of π-acceptor ligands such as carbon monoxide is to increase Δo. (see problem 7, page 175)

The next important point is how to describe the bonding for strong field ligands. Strong field ligands were actually discussed in Chapter 3 as *nonclassical ligands, π-bonding or π-acid -ligands.*

In Lewis acid/base theory, one can recall that a Lewis acid is an electron acceptor. *Nonclassical ligands are π-acid –ligands, and accept electron density from the metal by using appropriate π orbitals.*

Carbon monoxide (CO) is the prototype π-acceptor ligand. Carbon monoxide has a triple bond between the carbon and oxygen atoms, and these bonding molecular orbitals have corresponding *empty* anti-bonding π* molecular orbitals that can be utilized in bond formation with transition metals that have partially filled d-orbitals.

One of these can be drawn in the manner shown below:

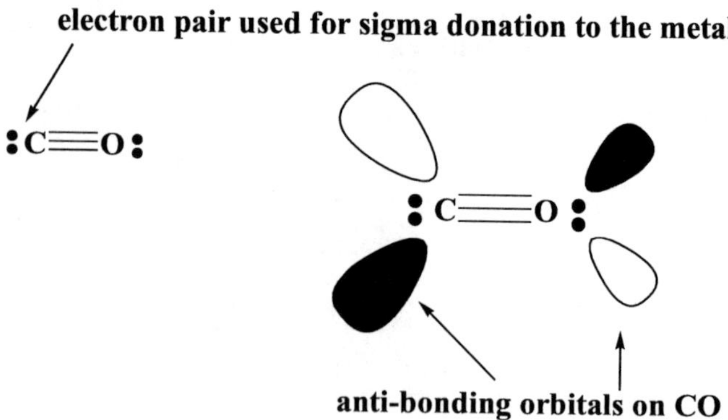

Carbon monoxide has three bonding orbitals that focus electron density mainly right between the carbon and oxygen atoms. The two anti-bonding orbitals (only one of which is shown in the diagram above) are empty, but have interesting overlap integrals that are strikingly like those of the d-orbitals.

If transition metals in the metal complexes have partially filled d-orbitals, then CO antibonding orbitals have the correct symmetry to merge with them. Carbon monoxide still possesses a lone pair of electrons on the carbon atom that can be donated to the metal. The π-bonding situation shown here is referred to as ***backbonding***.

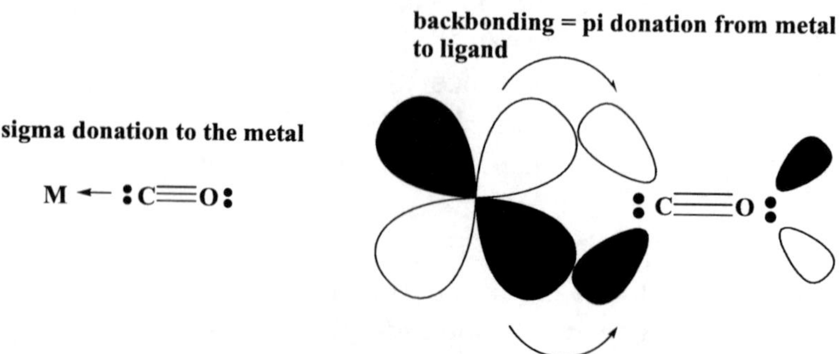

These pi bonds make up the t_{2g} set of molecular orbitals discussed on page 165 for the *π-acid ligands such as CO.*

Interestingly, this backbonding gives rise to two different bonding modes for CO to a transition metal.

The sigma donation is actually very weak since the lone pair on the carbon atom is held tightly (the lone pair on the oxygen atom is held even tighter---if you don't believe this calculate the Formal Charge on the oxygen atom) and thus the metal-carbon bond should be weak.

However, as we will see in the next chapter on organometallics, the metal-carbon bond is fairly strong, and metal carbonyl compounds are stable. The backbonding between the metal and the CO ligand, where the metal donates electron density to the CO ligand forms a ***dynamic synergism*** between the metal and ligand, which gives unusual stability to these compounds. This situation is shown below.

Metal -CO bonding is a Dynamic Synergism

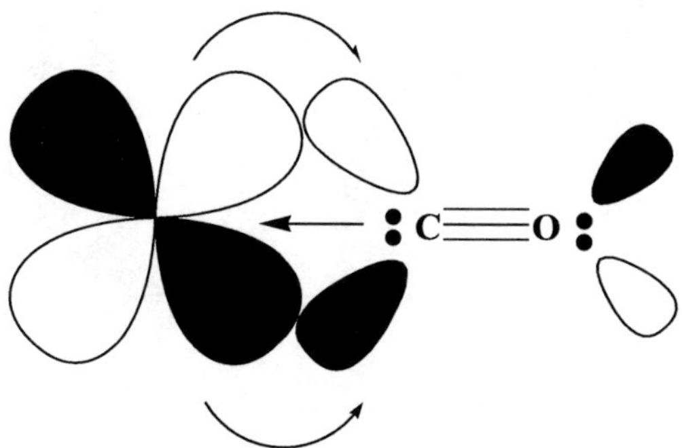

Utilizing Lewis electron dot structures and using a valence-bond approach (resonance forms) one could draw the structures and bonding like this:

A simple look at metal-carbonyl backbonding

$$M\!-\!C\!\equiv\!O: \quad\longleftrightarrow\quad M\!=\!C\!=\!O:$$

This is actually instructive because it suggests that the carbon-oxygen <u>bond order decreases</u> from 3 to 2 <u>as backbonding increases</u>. This can be followed easily using infrared spectroscopy since carbonyl (C=O) groups strongly absorb infrared radiation.

(stretching frequencies-see the discussion on **pages 178-179**)
Backbonding will be discussed further in Chapter 7.

The unoccupied carbon monoxide π-antibonding orbitals combine with the non-bonding t_{2g} orbitals (that are filled or partially filled with electrons) and become bonding. *This has the effect of lowering the t_{2g} orbital energies, and subsequently increasing Δo.* Thus, these type ligands (CO, CN^- and NO^+) promote a strong field.

Organophosphine Ligands

Phosphine ligands are strong field ligands also, but they exhibit backbonding in a quite different fashion than carbon monoxide. They achieve backbonding by using either their empty 3d orbitals or their empty sigma* orbitals. This issue has not been decided definitively and controversy exits as to which type of bonding shown here is actually the predominate one.

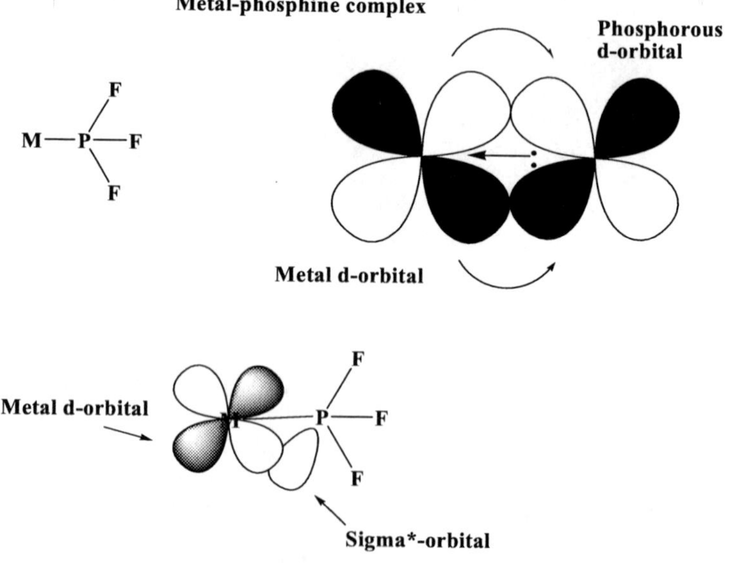

Interestingly, phosphine ligands can be modified so that they exhibit a wide range of properties. Two factors are most important: (1) the electronegativity (electron-withdrawing ability) of the substituents (the R groups), and also (2) their steric bulk. As the electronegativity of the R groups increases on the phosphine, the π-acceptor properties of the phosphine ligand increases.

Steric bulk can be measured by the *"cone angle"* of the phosphine ligand. Often, larger and bulkier phosphine ligands can promote active sites on the metal, which may increase catalytic activity. This is often seen with rhodium catalysts such as the ***Wilkinson's hydrogenation catalyst.***

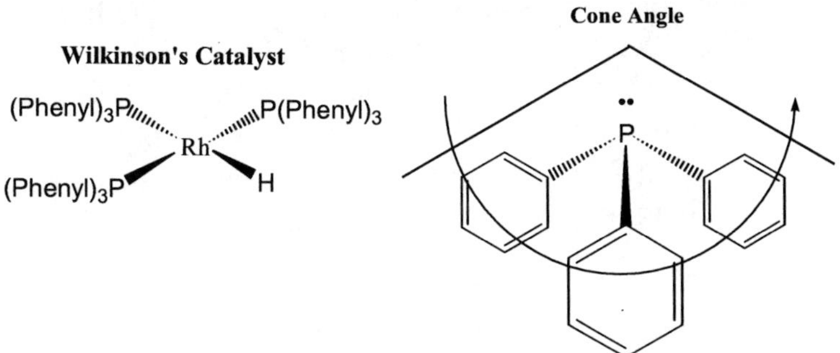

The organophosphine ligands can be synthesized so as to achieve a chiral phosphorous center (pages 89-90), which influences the hydrogenation reaction. This was the basis for the synthesis of L-Dopa using the chiral DIOP ligand.

Alkene Ligands

In 1827 Zeise prepared a platinum compound that he formulated as $KCl \cdot PtCl_2 \cdot EtOH$, but its structure was not established until the 1950's when it was shown to be a metal-alkene complex instead. The structure of the compound known as ***Zeise's Salt***, is shown below.

Some of the important features of this compound are that the alkene binds through both carbon atoms instead of only one atom, like most ligands, and that the alkene has a lowered carbon-carbon bond distance that suggests that the bond order is decreased from a double bond to somewhere intermediate between a double and single carbon-carbon bond.

The normal bond distance between the carbon atoms in a double bond is typically about 134 pm, and the bond distance between the carbon atom in Zeise's Salt is 137 pm. The typical carbon single bond distance is around 154 pm.

A similar compound is the amine complex trans-$PtCl_2(NMe_2H)(C_2H_4)$ which has a C-C bond distance of 147 pm, which is very close to that of a C-C single bond.

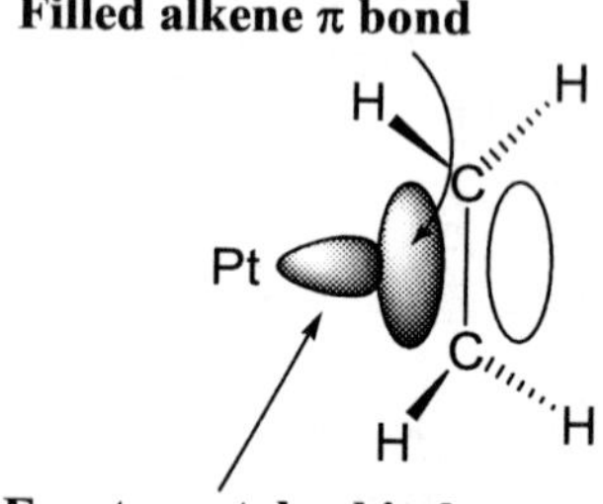

The bonding description is also very interesting, and the features of this bonding are shown below. Importantly, MO theory provides us with a framework in which to describe the backbonding in metal alkene complexes.

Sigma Bond Component   **Pi Acceptor Bond Component**

Filled alkene π bond   **Empty alkene π* orbital**

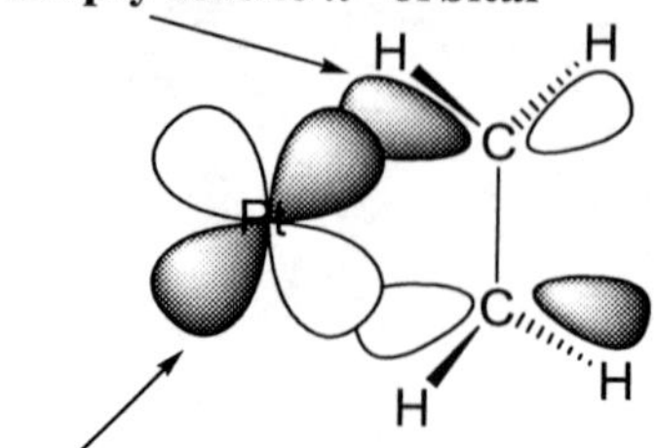

Empty metal orbital

The π-bond in the alkene acts as a sigma donor (like a lone pair) to form the sigma bond to the metal, and then the metal can participate in backbonding by pushing electron density in a d-orbital back onto the π^* antibonding orbital of the alkene, which reduces the bond order of the alkene.

Thus, the alkene acts as a σ-donor and a π-acceptor ligand.

Alkene-metal complexes are important for many commercial applications such as in the production of high-density polypropylene using *Ziegler-Natta* catalysts.

Summary: the Spectrochemical Series

After discussing the Ligand Field/Molecular Orbital treatment of the covalent bonding of ligands to transition metals, some useful observations emerge, and we are now in a position to explain the spectrochemical series.

The order of the ligands in the spectrochemical series is partly the strength with which they can participate in σ-bonding with the metal in question. More importantly however, is the participation and type of π-bonding that occurs between the ligand and the metal, for that is what determines the value of Δo.

When compared, using the common $e_g{}^*$ MO as our standard, then we see the differences in magnitude of Δo that the different ligands infer on metals in an octahedral ligand field. This comparison is shown below using only the MO's based on the d-orbitals.

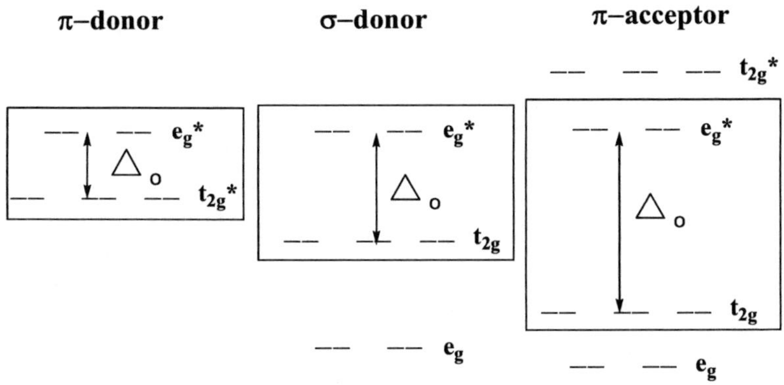

The overall order of the spectrochemical series may be rightly interpreted as being dominated by π effects.

The general series can then be seen as follows:

----------------------------Increasing Δo ------------------------$\rightarrow$

π-donor $<$weak π-donor$<$no π effects$<\pi$ acceptor

Examples of representative ligands for these various classes are:

    π-donor --- I^-
    weak π-donor ---Cl^-, F^-
    little or no π effects ---H_2O, NH_3
    π acceptor ---PR_3, CO

Terms and Definitions for Chapter 6

Sigma Bonding

Bond Order

Pi Bonding

Diatomic Molecules

Antibonding Orbital

Molecular Orbitals

Backbonding

Dynamic Synergism

Organophosphine Ligands

Zeigler-Natta Catalysts

Cone Angle

π-donor ligands

π-acceptor ligands

Alkene ligands

Wilkinson's Hydrogenation Catalyst

Linear Combination of Atomic Orbitals

Highest Occupied Molecular Orbitals (HOMO)

Lowest Unoccupied Molecular Orbitals (LUMO).

Chapter 6 Problems and Exercises

1. Use MO theory to determine the bond order in each of the two following species: $[He_2]^+$ and $[He_2]^{2+}$. Does the MO scheme indicate that these species have viable bonds?

2. Construct the MO diagram for O_2, but only show the 2-p valence orbitals involved. Next, use the diagram to *rationalize* the trend in oxygen-oxygen bond distances for the following species:
$O_2 \quad = 121$ pm
$[O_2]^+ = 121$ pm
$[O_2]^- = 121$ pm
$[O_2]^{2-} = 121$ pm
Which species is paramagnetic?

3. Draw the $\mathbf{a_{1g}}$, the $\mathbf{t_{1u}}$, and the $\mathbf{e_g}$ and $\mathbf{t_{2g}}$ sets of molecular orbitals that are expected for an octahedral metal complex. Make sure to show the overlaps between the metal and ligand based orbital portions.

4. Of the molecular orbitals drawn in problem #3 above, explain which are bonding and which are nonbonding.

5. Draw the scheme shown on page 163 that shows the MO diagram for a strong-field octahedral $[Co(NH_3)_6]^{3+}$ metal complex with a strong σ-donor ligand such as ammonia. Make sure that you label the $\mathbf{t_{2g}}$ set as non-bonding. Make sure that you draw the box around the HOMO and LUMO orbitals that define the energy spread for Δo.

6. Draw the scheme shown on page 164 that shows the MO diagram for an octahedral metal complex with a common π-donor ligand such as chloride. (Use $[CrCl_6]^{3-}$) Make sure that you draw the box around the HOMO and LUMO orbitals that define the energy spread for Δo.

7. Draw the scheme shown on page 175 that shows the MO diagram for an octahedral metal complex with a common π-donor ligand such as carbon monoxide. (Use [Cr(CO)$_6$]) Make sure that you draw the box around the HOMO and LUMO orbitals that define the energy spread for Δo.

8. Draw the synergistic bonding between a transition metal and a carbon monoxide ligand using the orbital overlap description.

9. Draw the synergistic bonding between a transition metal and a carbon monoxide ligand using the Lewis electron dot method showing the two resonance forms that are possible.

10. Why would iodide ion be lower on the spectrochemical series than the fluoride ion?

11. What kind of ligand would you expect the cyanide ion to be as far as sigma and pi interactions are concerned?

12. Why would water ion be lower on the spectrochemical series than the ammonia molecule? (hint—pi interactions)

13. Why would ammonia be lower on the spectrochemical series than the o-phenanthroline (phen) ligand.

14. Which would have the largest cone-angle… trimethyl phosphine or trimethyl phosphite?

15. Explain with drawings and discussion why Zeise's salt exhibits the structure with respect to the alkene ligand that it does.

Chapter 7.   Transition Metal Organometallics

The leading journal of the field, Organometallics, is an ACS publication and it uses this definition of "organometallics" in its Author Information section of its website.

"For the purposes of this journal, an 'organometallic' compound will be defined as one in which there is a bonding interaction (ionic or covalent, localized or delocalized) between one or more carbon atoms of an organic group or molecule and a main group, transition, lanthanide, or actinide metal atom (or atoms). Following longstanding tradition, organic derivatives of the metalloids (boron, silicon, germanium, arsenic, and tellurium) will be included in this definition. Furthermore, manuscripts dealing with metal-containing compounds that do not contain metal-carbon bonds will be considered as well if there is a close relationship between the subject matter and the principles and practice of organometallic chemistry. Such compounds may include, inter alia, representatives from the following classes: molecular metal hydrides; metal alkoxides, thiolates, amides, and phosphides; metal complexes containing organo-group 15 and 16 ligands; metal nitrosyls. Papers dealing with certain aspects of organophosphorus, organoselenium, and organosulfur chemistry also will be considered."

Organometallics is a huge area of chemistry that has come into its own in the last thirty years. Many organometallic compounds such as n-butyl lithium and the Grignard reagents (magnesium based) are essential building blocks in organic chemistry. Most organometallic compounds of the main-group metals are very reactive and many are extremely unstable. Grignard reagents are often air and water sensitive, and will easily decompose. Trivalent compounds such as AlR_3 and GaR_3 are electron-deficient (strong Lewis Acids) and are often *pyrophoric*.

Transition metal organometallics is a large field that is growing rapidly. It is an advanced topic, but it may be interesting to show a range of compounds that are found in this field. The metal carbonyls are found throughout the transition series, and the first row transition metals form these *binary metal carbonyls*:

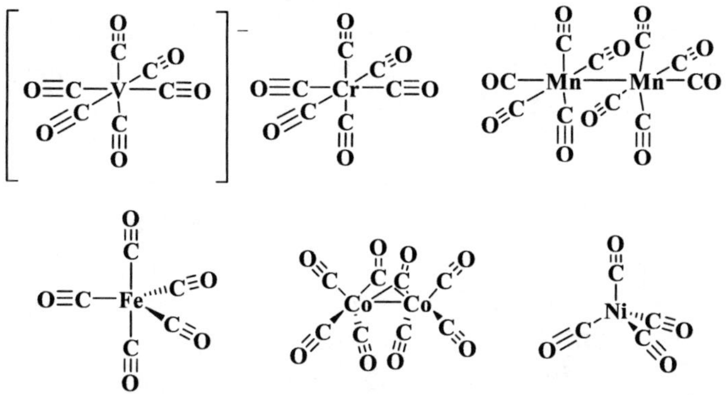

Binary metal carbonyls are also called *homoleptic compounds* since they *only have one type of ligand* around the metal.

The series shown above has an interesting trend. All of these compounds show a total of eighteen valence electrons around the metal center.

When a metal compound has eighteen electrons around the metal, it is said to obey the *eighteen-electron rule* (similar to the octet rule), which allows the metal to achieve a noble gas configuration. The eighteen-electron rule is also called the *noble gas formalism*, and the *Effective Atomic Number (EAN)* rule.

Another interesting note about this series of metal carbonyls is the bimetallic structures that are seen with manganese and cobalt. Since the two metals have an odd number of electrons, they cannot form neutral compounds with carbon monoxide alone unless they form metal-metal bonds. In this fashion they are able to achieve eighteen electrons around each metal.

You should count electrons for each compound to confirm that they each have eighteen electrons. A neutral vanadium carbonyl $[V(CO)_6]$ can't form an eighteen electron species because it would have only 17-electrons around the metal center, and it can't form a seven coordinate complex because of steric crowding. It is therefore very reactive, and it gains 18-electrons by becoming an anion $[V(CO)_6]^-$.

Electron counting will be discussed later in this chapter, and we'll work several examples. An example is $Cr(CO)_6$. Chromium is zero valent, and it therefore has a total of six valence electrons. Each carbonyl ligand donates two electrons for a total of twelve more electrons. Thus 6e⁻'s + 12 e⁻'s = 18 total.

Metal carbonyls are unusual in that the metal atom usually has a low oxidation state or has an oxidation state of zero. This is a consequence of the metal – carbonyl backbonding discussed in the last chapter, which stabilizes metal centers with larger numbers of d-electrons.

Carbon monoxide can also function as a bridging ligand, as can be seen for the dicobalt octacarbonyl compound (Co_2CO_8), or it can also bridge three metals.

Modes of carbonyl bonding are:

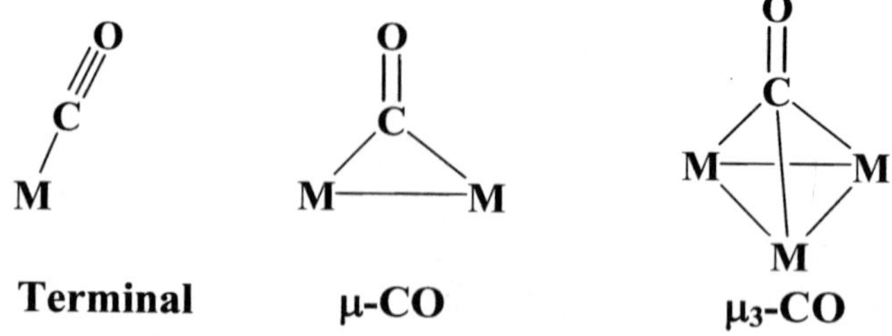

In the IR spectrum, free CO has a stretching mode at 2143 cm⁻¹. Absorption due to CO is strong and easily observed for metal carbonyls using infrared spectroscopy.

The lower the value of the stretching frequency, the weaker the CO bond, and conversely, the greater the back-bonding and the greater the metal-carbon bond.

The conclusions drawn from infrared data is supported by the structural data acquired from x-ray crystallography (and electron diffraction).

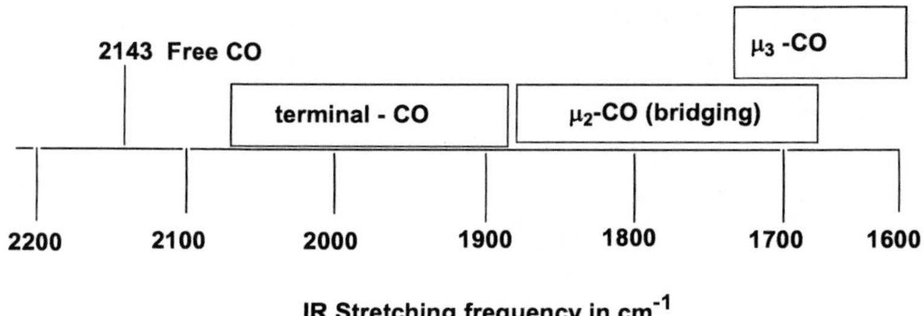

For free CO the carbon – oxygen bond length is 112.8 pm, whereas the terminal CO bond length in a metal carbonyl is typically 117 pm. A typical bridging CO bond length is longer, around 120 pm. <u>These longer bond lengths indicate stronger backbonding from the metal and a weaker C-O bond.</u> This experimental data strongly supports our molecular orbital approach to bonding for metal carbonyl compounds.

It is instructive to examine the data for an isoelectronic series of metal carbonyl species. For example, let's look at the following isoelectronic series of metal carbonyls and their CO stretching frequencies as well as their metal-carbon bond distances.

Complex	**γ-CO/ cm-1**	**M-C (pm)**
$Ni(CO)_4$	2060	184
$[Co(CO)_4]^-$	1890	175
$[Fe(CO)_4]^{2-}$	1790	174

The data shows that as more negative charge is placed on the metal center going from Ni^0 to Co^- to Fe^{2-}, then the backbonding increases. (This is because of the negative charge and the formal oxidation state of the metals, Ni^0, Co^{-1}, Fe^{-2}) This increased negative charge, and the subsequent greater backbonding, results in a lowering of the CO stretching frequency and a shortening of the metal-carbon bond.

Now, can you explain the data for the series above with respect to charge on the metal complex, backbonding, and the IR data?

Complex	γ-CO/ cm-1
$[Mn(CO)_6]^+$	2101
$Cr(CO)_6$	1981
$[V(CO)_6]^-$	1859

Metal carbonyls also have a tendency to form ***metal clusters*** that are nearly small chunks of metal surrounded by a carbon monoxide layer.

The metal clusters all exhibit metal-metal bonding. The metals share electron pairs, so in effect, each metal in a metal-metal bond is acting like a ligand to the other metal.

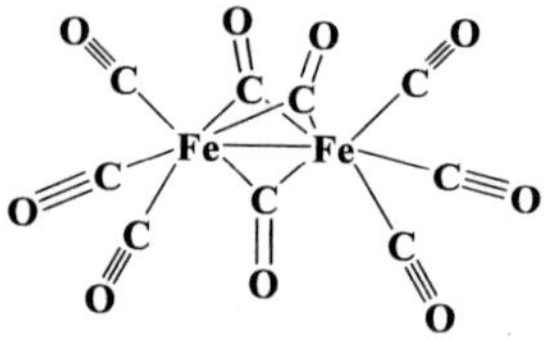

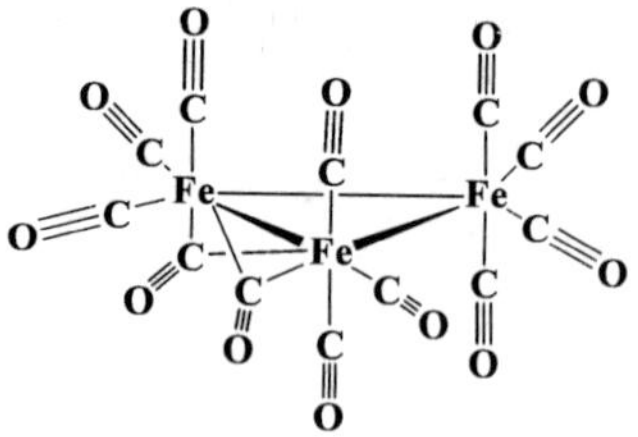

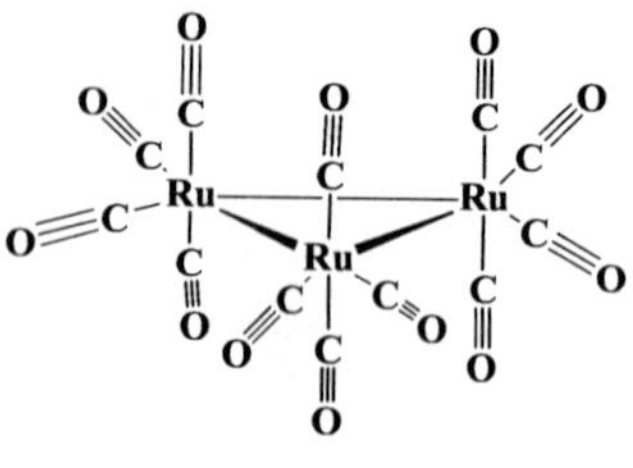

Fluxional Molecules

The metal carbonyls are often very ***fluxional molecules***, where the carbon monoxide ligands around the metal or metal centers are in motion and can vacillate between different geometries and can exchange between bridging and terminal sites.

An example of a geometric conversion has been discovered for iron pentacarbonyl, $Fe(CO)_5$, which undergoes what is called the ***Berry-Pseudorotation mechanism*** to convert between the trigonal bipyramid geometry and the square pyramidal geometry. This mechanism allows the CO ligands at the two different sites of the trigonal bipyramid (the axial and equatorial sites---see Chapter 2) to interconvert on the NMR time scale.

At room temperature the ^{13}C NMR shows only one peak, one kind of carbonyl, because the interconversion is so fast, but at colder temperatures, as seen with variable temperature NMR, the two different carbonyl ligands can be observed.

Metal carbonyl dimers and clusters often exhibit a fluxional process where terminal and bridging carbonyls interchange more rapidly with increasing temperature.

A commonly used example is that of dicobalt octacarbonyl, $Co_2(CO)_8$, which exhibits both a bridged form and an all-terminal form.

$Mn_2(CO)_{10}$ prefers the staggered form.

Metallocenes

Ferrocene was an important organometallic compound discovered in the 1950's due to the fact that it showed for the first time the synthesis of a metal-arene complex:

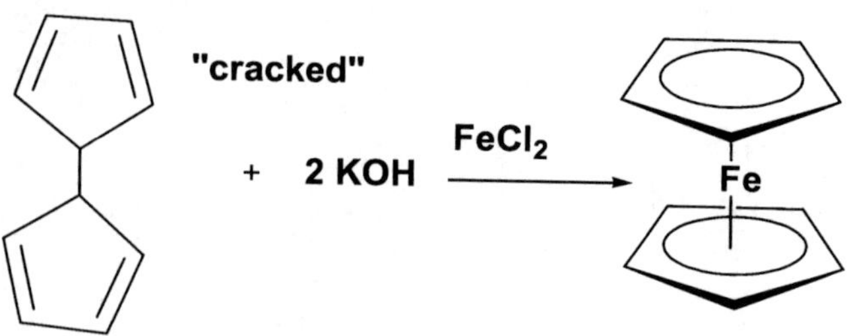

The cyclopentadienyl ring (cp⁻) is an anion and is aromatic (with six delocalized electrons). In electron counting, does ferrocene satisfy the eighteen-electron rule?

The cp⁻ rings are each six-electron donors giving a total of twelve electrons for the ligands. The iron atom has eight valence electrons, but since the molecule is neutral with no charge, the iron has to be formally in the Fe^{+2} oxidation state. Thus, it has only six electrons. When added up, ferrocene has eighteen electrons around the iron center.

Many metal arene complexes have been synthesized since that time, including many ***sandwich*** type molecules similar to ferrocene.

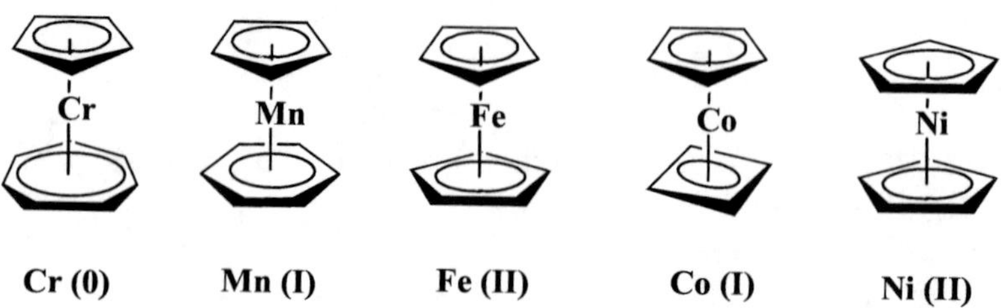

In a sandwich complex, the metal lies between two arene ligands. Complexes of the general type $(\eta^5\text{-Cp})_2\text{M}$ are called ***metallocenes***. (metallocenes do not *have* to be η^5-Cp) Can you determine if the metallocenes depicted above all obey the eighteen-electron rule? (see problem 17)

Electron Counting Revisited

Even though we covered two electron-counting methods in Chapter 5, the needs of transition metal organometallic chemistry are such that it is important to revisit the more important of the two methods, the **ICC** method. One of the simplest cases is for the binary metal carbonyl, chromium hexacarbonyl.

$$Cr(CO)_6$$

Cr	6e-
6 terminal CO	12e-
	18e-

This electron counting is done for the chromium center, and therefore the compound obeys the 18-electron rule. The formal oxidation state of chromium in this case is Cr^o.

A more complicated case arises when a metal-metal bond is present, such as in the dimeric binary metal carbonyl compound, dimanganese decacarbonyl.

$$Mn_2(CO)_{10}$$

Mn	7e-
M-M bond	1e-
5 terminal CO	10e-
	18e-

In this case the metal-metal bond plays an important role. The ·Mn(CO)$_5$ fragment acts as a ligand for the metal being counted, and becomes a one electron donor ligand through sharing. Each metal center has the same number of electrons: 7e⁻ from the metal, 1e⁻ from the other metal in the metal-metal bond, and 10e⁻ from the five terminal carbonyl ligands.

The metal is formally in the Mn^o oxidation state; the other Mn atom does not oxidize it.

Another example of a dimeric binary metal carbonyl is dicobalt octacarbonyl.

$Co_2(CO)_8$

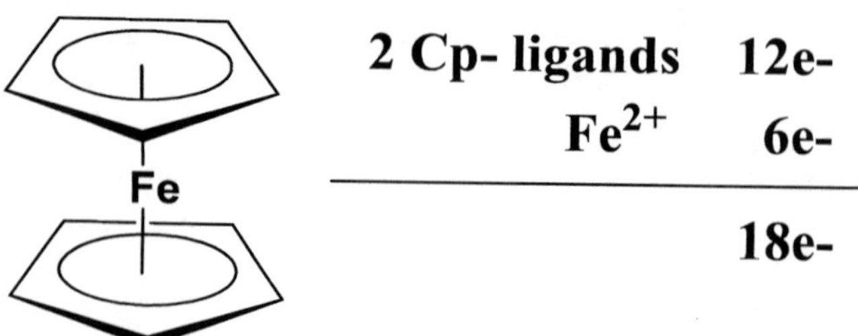

Co	9e-
M-M bond	1e-
3 terminal CO	6e-
2 bridging CO	2e-
	18e-

The difference between these two comes from the appearance of the bridging carbonyl binding mode. Since the carbon monoxide ligand is only using a pair of electrons to bond to the two metal centers, then, by convention, we assign only one electron to each metal. Again, in this case, the metal is formally in the Co^o oxidation state; the other Co atom does not oxidize it, nor does the bridging carbonyls.

Continuing on, let's examine ferrocene. Ferrocene has no carbon monoxide ligands, but is synthesized by using an in-situ production of cyclopentadienyl anion (Cp^-).

Ferrocene

2 Cp- ligands	12e-
Fe^{2+}	6e-
	18e-

For ferrocene, it is best to start with the ligands and remove them with their full valence shell electrons. Thus, we remove the ligand as Cp^-, which is aromatic. The Cp^- anion is therefore a six-electron donor ligand, and this results in an oxidation of iron by plus one for each Cp⁻ ring. Even though iron is formally in the Fe^{+2} oxidation state in ferrocene, it must be realized that this is a convention only. Ferrocene is non-polar, dissolves in organic solvents, and does not behave like an ionic compound. It is a molecular compound.

The following compound is similar to ferrocene, and can be synthesized from it, but it has a benzene ring as a ligand.

$$[(\eta^6-C_6H_6)(\eta^5-C_5H_5)Fe]^+$$

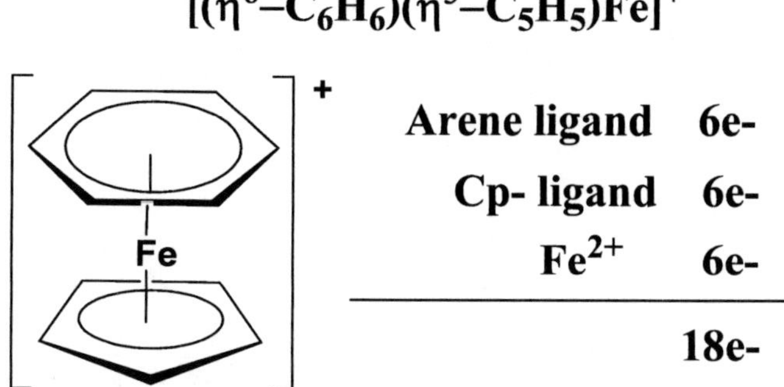

The complex has a Cp⁻ ring which is a six electron donor and a benzene (arene) ring which donates six-electrons. The overall charge on the complex is 1^+, and since the Cp⁻ ring is negatively charged, the iron atom has to have an oxidation state of Fe^{+2}. Thus, the compound obeys the 18-electron rule.

Cycloheptatriene acts as an aromatic 6-electron donor by loss of a hydride (H⁻) to form the tropylium ion.

Cycloheptatriene

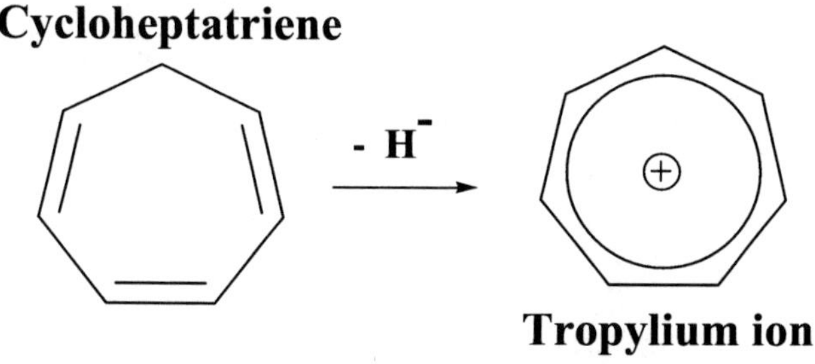

Tropylium ion

The *tropylium ion* reacts with transition metals in a similar fashion to benzene or cyclopentadienyl anion, except that it has a positive (1^+) charge.

This metal complex has an overall charge of 1^+, so the molybdenum has a formal oxidation state of Mo^o.

$$[(\eta^7-C_7H_7)Mo(CO)_3]^+$$

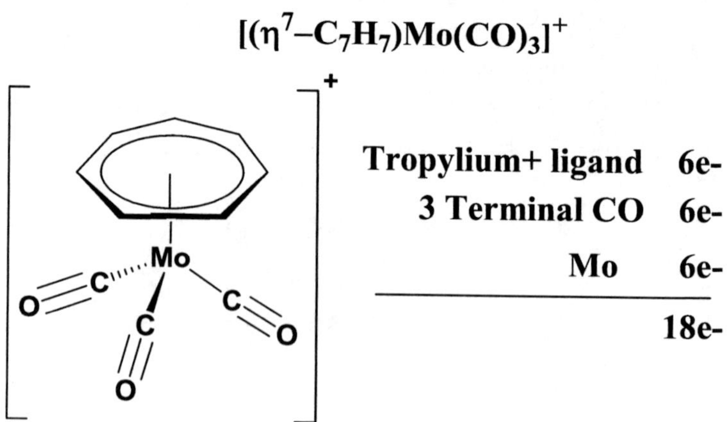

Tropylium+ ligand	**6e-**
3 Terminal CO	**6e-**
Mo	**6e-**
	18e-

The cyclobutadiene molecule is not aromatic. However, in a beautiful twist of nature, cyclobutadiene can be made aromatic-where all the bond lengths are the same-by reacting it with a transition metal to form an organometallic compound.

In this case, the ligand is a four electron neutral donor ligand.

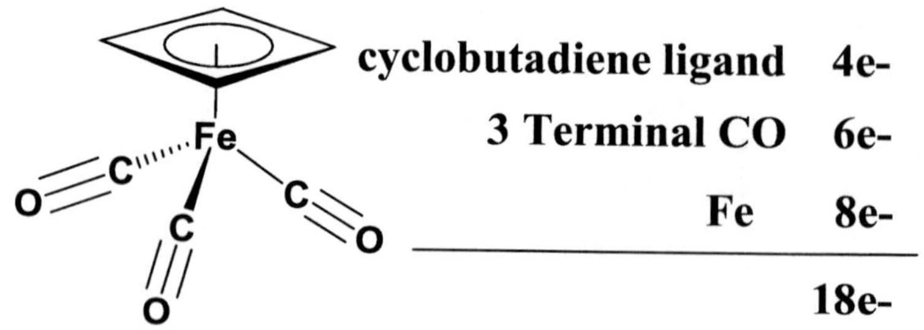

cyclobutadiene ligand	**4e-**
3 Terminal CO	**6e-**
Fe	**8e-**
	18e-

Vaska's complex undergoes oxidative addition to form six-coordinate octahedral complexes, some of which contain stable alkyl metal bonds.

triphenylphosphine ligands	4e-
Terminal CO	2e-
CH_3^-	2e-
Cl^-	2e-
I^-	2e-
Ir^{3+}	6e-
	18e-

In this case the triphenyl phosphine ligands are neutral, but the chloride, the iodide and the methyl groups are all anionic ligands. These three anionic ligands oxidize the iridium to Ir^{3+}, so that is the formal oxidation state of the metal.

The nitric oxide ligand is isoelectronic with carbon monoxide, but the major difference is that has a *positive charge*.

4 Terminal CO	8e-
Terminal NO^+	2e-
Mn^-	8e-
	18e-

The interesting feature of the nitrosyl ligand is that it *decreases* the oxidation state of the metal. Thus, the manganese metal center has an oxidation state of <u>minus one</u> (Mn^-).

Grubb's first commercial catalyst is very useful in organic chemistry due to its usefulness as a metathesis catalyst. To count the electrons one must pull off the carbene in its closed shell configuration. This would make the carbene an anion with a 2^- charge. Thus, formally, the Ru would have to be in the Ru^{4+} oxidation state, and the trigonal bipyramidal complex would be a sixteen electron species.

Grubb's Catalyst

(1st commercial catalyst)

Ru^{4+}	4e-
2 Cl-	4e-
carbene^{2-}	4e-
2 tricyclohexylphosphine ligands	4e-
	16e-

Oxidative Addition and Reductive Elimination Reactions

A variety of industrially important catalytic cycles have as their basis a series of reactions involving an *oxidative addition reaction*, a rearrangement and then a *reductive elimination reaction*.

What is an oxidative addition reaction? An oxidative addition reaction is as the name suggests---a ligand (a general example is ligand A-B) reacts with a metal complex that is *coordinatively unsaturated* in such a manner that the metal is formally oxidized by +2 (ICC method of electron counting). An excellent example can be made by using one of the classical metal complexes, called *Vaska's complex*, in an oxidative addition reaction with HBr. Vaska's complex is a square-planar iridium(I) complex that is, of course, coordinatively unsaturated.

Oxidative Addition Reaction

(Ph)$_3$P — Cl — Ir — C≡O — P(Ph)$_3$ Vaska's Complex

HBr →

H — (Ph)$_3$P — Br — Ir — C≡O — P(Ph)$_3$ — Cl

Ir(I)   Ir(III)

Square-Planar 16 e⁻ species   Octahedral 18 e⁻ species

Oxidative addition reactions are more likely to occur in metals with low oxidation states compared to those with a higher oxidation state. Thus, a Rh(I) species is more likely to undergo an oxidative addition reaction compared to a Rh(III) species.

Oxidative addition reactions cannot occur with metals in their highest oxidation state. For example, a Mn(VII) species such as MnO_4^- cannot undergo an oxidative addition reaction.

Ligand species that have a weaker A-B bond are more likely to undergo an oxidative addition reaction compared to those with a stronger A-B bond. Thus, a C-Cl ligand bond is more likely to undergo oxidative addition as compared to a C-C ligand bond.

A reductive elimination reaction is the microscopic reverse reaction of an oxidative addition reaction. In the reductive elimination reaction the ligands that are eliminated *must be cis* to each other. An example is given below, where the iridium(III) complex (coordinatively saturated) eliminates acetone as a product, and iridium is formally reduced to Ir(I), and becomes coordinatively unsaturated.

Reductive Elimination Reaction

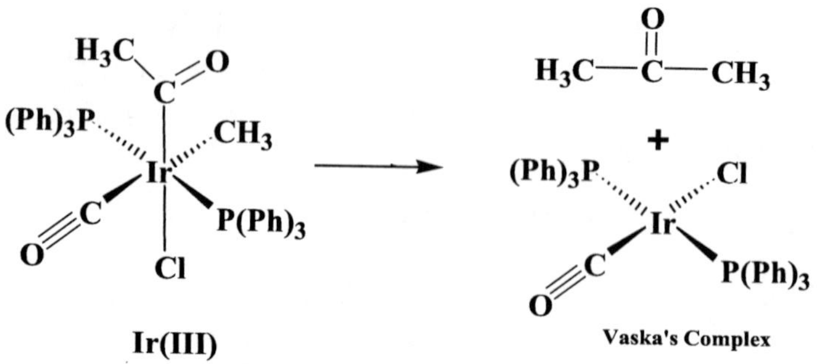

Ir(III)

Octahedral 18 e⁻ species

Ir(I)

Square-Planar 16 e⁻ species

The reductive elimination step is a crucial step in many industrial catalytic processes (which explains the common use of rhodium) such as the Monsanto acetic acid process, the Tennessee Eastman acetic anhydride process, and in many hydrogenation reactions.

Metal-Metal Multiple Bonds

The existence of a *quadruple metal-metal bond* in inorganic compounds was first recognized in 1964 when the compound $[Re_2Cl_8]^{2-}$ was isolated.

Unfortunately, the complex was actually first synthesized ten years prior to this in the Soviet Union, but was characterized mistakenly as the Re(II) compound, K_2ReCl_4.

In a now classical paper, F. Albert Cotton assigned the correct formula and structure as being a one that contained a Re-Re quadruple bond due to the odd eclipsed structure uncovered by x-ray crystal analysis.

The compound consists of two d^4 $ReCl_4$ groups, which contain Re(III) metal centers, and a metal-metal bond distance of only 223 pm.

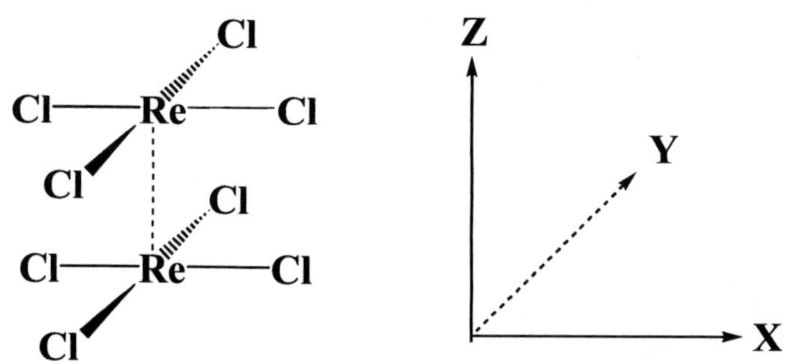

The scheme above shows that the exhibited geometry is that of two square-planar $ReCl_4$ units stacked on top of one another to form an eclipsed structure where the eight Cl ligands form a nearly perfect cube.

From an electrostatic and steric point of view, this compound should exhibit the staggered conformation.

Since the staggered conformation is not observed, then there must be unexplained bonding considerations. The bonding can most easily be explained by considering the space orientation of the d orbitals. Each rhenium is slightly displaced above the center of a square planar array of four chloride ions. The metal dx^2-y^2 orbital has the appropriate symmetry to bond to the four chlorides.

The 1st bond between the metals uses the dz orbitals on each metal, to form a σ bonding molecular orbital.

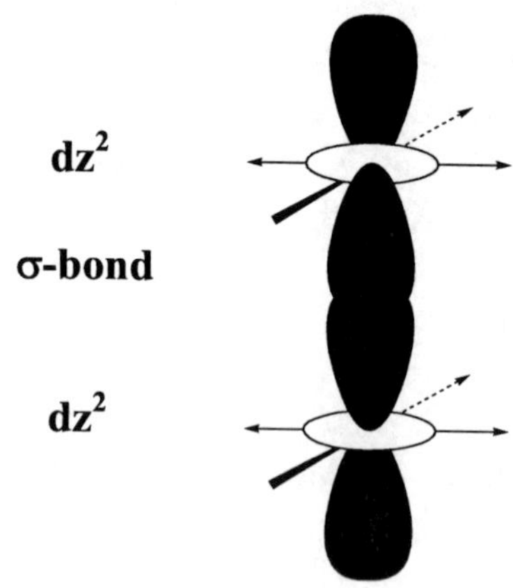

The 2nd and 3rd bonds between the metals use the dxz and the dyz orbitals on each metal, to form two π bonding molecular orbitals. Both π -bonding MO's are shown here.

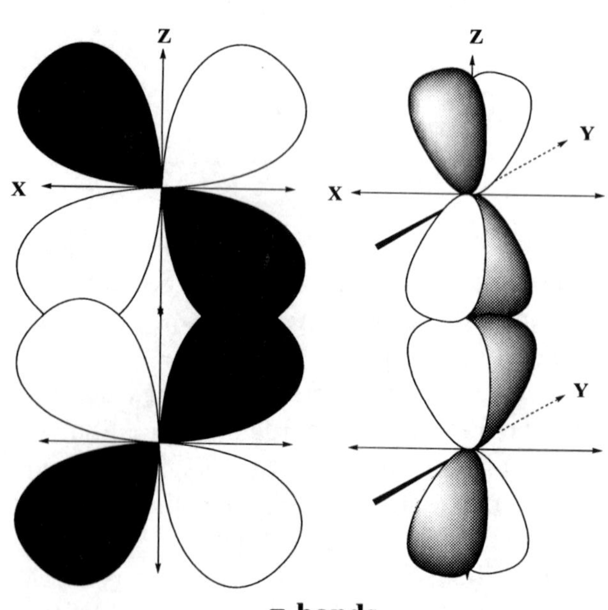

The 4th bond between the two metals is formed by the two dxy orbitals that are parallel to each other and form a type of bond not seen in organic chemistry called a ***δ (delta) bond***. The "delta", with a "d", name comes from the fact that it looks like a d-orbital-shaped bond. The dx^2-y^2 orbital is used to form M-L bonds and is not involved in the M-M bond.

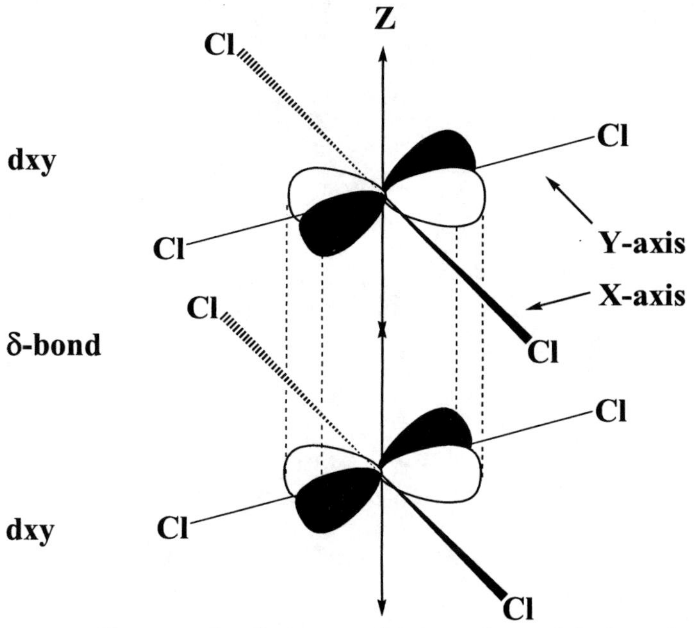

For overlap between the two dxy orbitals to be maximized in order to become a bonding interaction, the two ReCl$_4$ square planes <u>must be eclipsed</u> to each other. Despite the fact that this maximizes inter-atomic repulsions between the chloride ligands, the ability of the two metals to quadruple bond is the overriding stabilizing factor.

Perhaps the most interesting feature of this compound is the δ interaction in the quadruple bond. Because the δ orbital is only weakly bonding and the δ* orbital is only weakly antibonding a number of interesting chemical and spectroscopic consequences result. For example, the brilliant blue color of [Re$_2$Cl$_8$]$^{2-}$ is due to a δ----> δ* electronic transition. Because of the weakness of the δ bond, the gain or loss of electrons has a relatively minor effect on the strength of the M-M bond.

The following scheme shows the summation of the bonds in the quadruple bond, with the delta bond being the highest energy molecular orbital.

dxy   **δ-bond**

dxz, dyz   **π-bonds**

dz^2   **σ-bond**

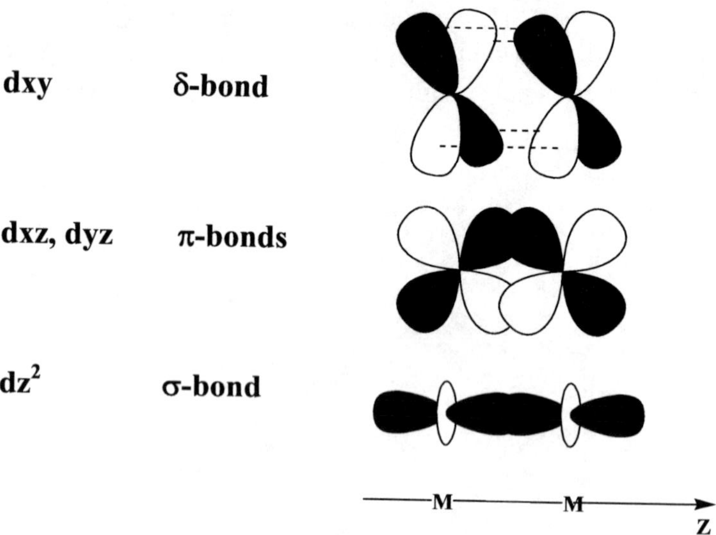

The MO energy diagram is shown below. The delta, pi, and sigma bonds have filled orbitals, but the antibonding orbitals are unfilled.

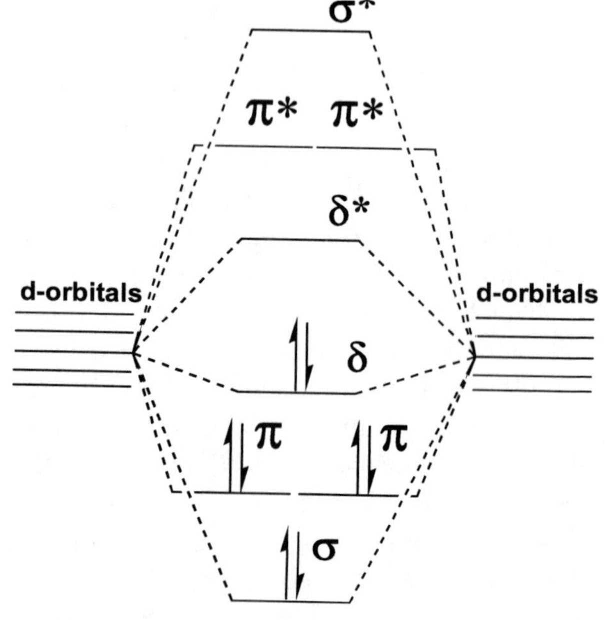

Terms and Definitions for Chapter 7:

Pyrophoric

Organometallics

Binary Metal Carbonyls

Homoleptic Compounds

Eighteen Electron Rule
Noble Gas Formalism
Effective Atomic Number (EAN)

Metal Clusters

Fluxional Molecules

Berry-Pseudorotation Mechanism

Ferrocene

Sandwich Type Molecules

Ziegler-Natta Polymerization

Oxidative Addition Reaction
Reductive Elimination Reaction

coordinatively unsaturated

Vaska's complex

Quadruple Metal-Metal Bond

δ (delta) bond

Tropylium Ion

Chapter 7 Problems and Exercises

1. Write the formulas for the neutral mononuclear metal carbonyl molecules formed by V, Cr, Fe, and Ni. Which ones satisfy the noble gas formalism?

2. Why are the simplest carbonyls of the metals Mn, Tc, Re, and Co, Rh, and Ir, polynuclear?

3. Explain, with necessary orbital diagrams, how CO, which has negligible donor properties toward simple acceptors such as BF_3 can form strong bonds to transition metal atoms.

4. In what ways can CO be bound to a metal atom?

5. Discuss and explain the trend in CO stretching frequencies in the series $V(CO)_6^-$, $Cr(CO)_6$, $Mn(CO)_6^+$.

6. Draw the structures of
    (a) $Fe_2(CO)_9$,
    (b) $Ru_3(CO)_{12}$
    (c) $Rh_4(CO)_{12}$

7. What kind of ligand would hydrogen be counted as in a transition metal complex? Count the electrons for hydridocobalt tetracarbonyl $[HCo(CO)_4]$ and verify that it obeys the 18-electron rule. What is the oxidation state of the cobalt?
What if you discover that the compound is acidic in water, would that change your electron counting formalism?

8. Confirm that the iron center in $H_2Fe(CO)_4$ obeys the 18-electron rule. What is the formal oxidation state of iron when using the ICC electron counting technique?

9. Confirm that the manganese center in $HMn(CO)_3(PPh_3)_2$ obeys the 18-electron rule. What is the formal oxidation state of manganese when using the ICC electron counting technique?

10. By using the 18-electron rule show that the zero-valent metal carbonyls of the manganese Group (Mn, Tc, Re) should all exhibit a metal-metal bond in their $M_2(CO)_{10}$ compounds.

11. Confirm that the metal centers all obey the 18-electron rule in the following complexes. Some do not and have 16-electrons instead. Draw the structures.
    (a) cis-$Mo(CO)_4(pamp)_2$
    (b) $Fe(CO)_5$
    (c) di-μ-carbonylbis(tricarbonylrhodium)(0)
    (d) $W(CO)_6$
    (e) $Mo(CO)_4$dipamp
    (f) $(\eta^5\text{-Cp})_2Fe$
    (g) $[RhH(triphenylphosphine)_3]$
    (h) $[Ir(CH_3)H(CO)Cl(triphenylphosphine)_2]$
    (i) $^{99m}[Tc(dmpe)_3]^+$
    (j) $(\eta^5\text{-Cp})Tc(CO)_3$
    (k) $(\eta^6\text{-}C_6H_6)_2Cr$
    (l) $(\eta^5\text{-Cp})(CO)Fe(\mu\text{-CO})_2Fe(CO)(\eta^5\text{-Cp})$ this Homobimetallic compound contains a metal-metal single bond

12. Draw the Berry Pseudorotation mechanism for $Ru(CO)_5$ that makes all the carbonyl ligands interchange on the ^{13}C NMR timescale at room temperature and become equivalent.

13. The M-CO bonding is synergetic or involves synergism.
a. What is synergism?
b. Explain the following trend in the magnitude of $\nu_{C\text{-}O}$ IR frequencies:

free CO $>$ terminal M-CO $> (\mu^2\text{-CO})M_2 > (\mu^3\text{-CO})M_3$

14. For the following organometallic dimer complexes, determine the metal-metal bond order assuming that each complex obeys the 18-electron rule. Draw the structure of each complex and show how you counted the 18 electrons next to each structure.

(a) di-μ-carbonylbis(tricarbonylcobalt)(0)

(b) $[(\eta^5\text{-}C_5H_5)Mn(CO)_2]_2$

(c) $[(\eta^5\text{-}C_5H_5)Mo(\mu\text{-}CO)_2]_2$

15. Draw the structure of the following compound and predict the bond order of any metal-metal bond in the $[Re_2Cl_6(\mu\text{-}DIPHOS]$ dimer.

16. Count the electrons for the metal sandwich compounds found on the cover, and verify that the oxidation states are correctly listed.

17. Draw the square planar $[RuCl(NO)(PPh_3)_2]$ complex. How many isomers are present? Verify that every isomer has sixteen or eighteen electrons around the metal center. What is the formal oxidation state of the Ru metal center?

18. Draw the tetrahedral $[Co(CO)_3(NO)]$ complex. Verify that it has sixteen or eighteen electrons around the metal center. What is the formal oxidation state of the Co metal center?

19. Draw the tetrahedral $[Fe(CO)_2(NO)_2]$ complex. Verify that it has sixteen or eighteen electrons around the metal center. What is the formal oxidation state of the Fe metal center?

20. Confirm that each different metal center in the following compound obeys the 18-electron rule. $[Fe_3(CO)_{12}]$

Chapter 8. Inorganic Laboratory Experiments

These lab experiments are designed for a senior level inorganic laboratory, and they focus on the synthesis and characterization of transition metal complexes that are used for specific purposes in the field of *Nuclear Medicine* in some fashion. This chapter consists of two experiments that are based on the syntheses of the "cold" non-radioactive forms of currently used *radiopharmaceuticals* in the medical field.

Experiment 1: Synthesis of Cu(PTSM)

Experiment 2: Synthesis Of $N(Butyl)_4^+[ReO(DMSA)_2]^-$.

In keeping with the thrust of this book---Coordination Chemistry--- these compounds exhibit interesting coordination geometries using not-so-common ligands that interact in a fascinating way with the metal of interest.

Experiment 1: Synthesis of Cu(PTSM)

Introduction

The synthesis of multidentate ligands has been of great research interest in the last twenty years due to the great stability of the coordination complexes they can form. Metal complexes containing chelate rings have an enhanced stability that is termed the chelate effect.

Many new ligands must be synthesized on-site since these specialized ligands are not as readily available as many common organic compounds or organic pharmaceuticals.

In this experiment we must first prepare a tetradentate ligand (PTSM) named pyruvaldehyde dithiosemicarbazone.

The compound is synthesized easily and directly from the condensation reaction of pyruvaldehyde with 4-methyl-3- thiosemicarbazide. This is depicted in the next scheme.

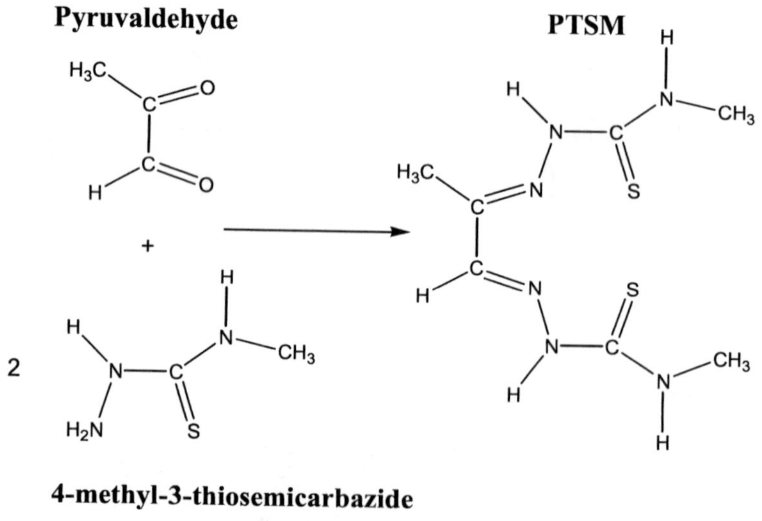

The reaction of the PTSM ligand with metal ions can form interesting metal complexes. This is especially true with the Cu^{2+} ion, for which it has a great affinity.

The reaction of PTSM with the copper II ion is shown in the next scheme.

PTSM Cu^{2+} →   **Cu-PTSM** $+ 2\,H^+$

In this case, the Cu-PTSM complex is a highly colored red species, thus its formation can be used as a way to detect aqueous Cu^{2+} ions at very dilute concentrations. The ligand itself has free rotation around some bonds and wraps around the copper ion, releasing two protons during the coordination process. This has the effect of placing a negative charge on the sulfur atoms, which balances out the 2+ charge on the copper ion.

The resulting molecule is neutral and has no charge.

The Cu-PTSM complex is medically attractive when synthesized with the positron-emitting isotope Cu-64, which has a half-life of 12.9 hrs. The Cu-64 isotope, as well as the Cu-62 isotope, is a positron-emitting isotope, which by annihilation of the positron produces two gamma rays of 512 KeV energy at an angle of 180 degrees from each other. This makes it easy for a PET scanner in a clinical setting to detect, to the fraction of a millimeter, exactly where the isotope was when it decayed.

The Cu-64 PTSM complex is used in brain, heart, and blood scans to detect diseased tissue and organ malfunctions. The ease of preparation of the complex makes it attractive in a clinical setting.

In the clinical setting the ligand is dissolved in ethanol with DMSO, and then reacted with the radioactive

Cu-64 isotope as the chloride or nitrate salt. The extremely dilute solution of radioactive product, which keeps the complex in solution, has an excess of ligand. It is filtered and injected as prepared.

Chemicals Needed

(a) 4-methyl-3-thiosemicarbazide
(b) pyruvaldehyde
(c) copper acetate dihydrate
(d) 5% acetic acid
(e) ethanol
(f) methylene chloride
(g) magnesium sulfate
(h) methanol

Procedure

A 25 mL flask is charged with 0.010 mol (________g) of 4-methyl-3-thiosemicarbazide. This is slurried in 15 mL of 5% acetic acid. The slurry is heated to 60 degrees C, and then to the slurry 0.004 mol (________ml) of pyruvaldehyde in 3 ml of ethanol or methanol is added dropwise over 5-10 minutes. After another 15 minutes of stirring at 60 degrees, the reaction mixture is allowed to cool, and then filtered. The ligand can then be recrystallized from 50/50 methanol/water. Acquire the melting point. (217-219)

Dissolve 5.0×10^{-5} mol of Cu(acetate) dihydrate in 1 mL of water and then test the pH. Next, add 4.0×10^{-5} mol of the ligand dissolved in 1 mL of methylene chloride (dichloromethane) to the vial containing the copper solution. Is there a reaction? How can you tell? Are there two phases present? Next add 0.5 mL of ethanol. Allow the reaction to proceed for 30 minutes to an hour while stirring. Test the pH of the aqueous layer. Separate the product layer by adding up to 5ml of water and 10ml of methylene chloride (which is more dense, water or methylene chloride?) and dry the separated organic layer that contains the product with a spatula tip of magnesium sulfate.

Take a TLC (thin layer chromatography) of the red product using a silica plate with 50/50 ethyl acetate/ hexane as the mobile phase (the mobile phase may need to be modified, using an adjusted ratio of solvents!). Use only a small spot on the plate of the product containing solution. How pure does the complex appear to be? Does the product run with the free ligand? Does the product smear on the plate? If so, can you give a possible explanation? Let the product dry by evaporation of the solvent, dry the crude product, and then get the percent yield. Is the product an ionic compound or nonpolar? How do you know? Is the complex tetrahedral or square planar?

Get the UV-Vis spectrum of the ligand (0.00007 M) 1 and the metal complex (0.00003 M) 2 and find the lambda max and the molar absorptivity of each. Acquire the ^{1}H NMR on the ligand and the thiosemicarbazide starting material. Compound 1 is the ligand, and compound 2 is the copper complex. Try to run an NMR on the copper complex. Any luck?

Sample results:

Table 1. The ^{1}H NMR data for the TSC ligand (in ppm).

Compound	NH-CH$_3$	C-NH-C	-N-NH
1	3.0(d)	8.4(m), 8.6 (m)	10.4(s), 11.8(s)

All samples were 10 mg of each ligand in~ 1 ml of d$_6$-DMSO with TMS internal standard.

Table 2. Spectroscopy data for the TSC ligands and the Cu(II) complexes.

Compnd	M(mol/L)	λmax	Abs	ε (molar abs.)
1	0.000073	360.4	3.0093	4122
2	0.000036	358.8	3.1197	8666

References

(1) Jurrison S., Berning D., Jia W., Ma D.-Chem . Rev. 93: 1137 (1993)

(2) Green M. A., Klipperstein D. L., and Tennison J. R. - J. Nucl. Med. 29: 1549 (1988)

(4) Wada K., Fujibayashia Y., Yokoyama A. - Arch. Biochem. and Biophys., 310: 15 (1994).

(5) McPherson D. W., Umbricht G., Knapp F. F. - J. Labelled Compd. Radiopharm. 28: 877 (1990)

(6) Taylor M. R. - Abstracts of the American Crystallographic Association. Gatlinburg, Tennessee, 1965, Abstract D-4.

(7) Petering H. G., Buskirk H. H., Underwood G. E. - Cancer Res. 24: 367, (1964)

(8) Coates E. A., Milstein S. R., Holbein G., McDonald J., Reed R., and Petering H. G. - J. Med. Chem. 19: 131 (1976)

(9) John E. K., Green M. A. - J. Med. Chem. 33: 1764 (1990)

Experiment 2: Synthesis Of N(Butyl)$_4$$^+$[ReO(DMSA)$_2$]$^-$.

Introduction

Many transition metal compounds are formed by using suitable starting materials where the transition metal is already in the correct oxidation state, and ligands are simply substituted onto the metal center or core.

For many rare transition metals, such as the heavier 2nd and 3rd row transition metals, the metals can only be readily obtained as metal salts, such as $RhCl_3$, or as oxides, such as WO_3. Often, purified metals are dissolved in mineral acids to obtain these metal salts or oxides. In cases where the transition metals can't be obtained in the desired oxidation state, reagents that give selective reductions or oxidations must be used, as well as appropriate choices of ligands.

Rhenium is a rare and expensive metal that has an extensive coordination chemistry and a range of eight oxidation states from Re(VII) to Re(-I). Because of its scarcity, only a few starting compounds for making Re complexes are commercially available. One of these starting materials is the perrhenate ion (ReO_4^-).

Interest in Re compounds has developed quickly in the last 10-20 years due to the availability of radioisotopes of Re (^{186}Re and ^{188}Re) which are useful for therapeutic nuclear medicine procedures. The isotopes are available only in the chemical form of the perrhenate ion. The perrhenate ion is of no use as a radiopharmaceutical itself because it is excreted by the body over a 24-hour period of time through the urine and feces, and does not localize in tissues of interest such as tumors and bone.

This experiment deals with the synthesis of a "cold" Re compound---ReO(DMSA)$_2$$^-$ ---on a milligram scale. ("Cold"= nonradioactive). The synthesis involves the reduction of Re(VII) to Re(V) by use of a mild reducing agent, in this case the stannous ion Sn^{2+}. Stannous ion itself is oxidized to SnO_2, which precipitates out of solution.

This Re compound is of interest to nuclear medicine clinicians because it (and the Tc analog) accumulates preferentially in some human tumors, especially medullary thyroid carcinoma. Since ^{99m}Tc is a γ (gamma) emitting isotope, and the ^{188}Re and 186Re isotopes are both primarily β (beta) emitting isotopes, then the possibility of a "matched pair " of diagnostic (Tc) and therapeutic (Re) agents for these cancers arises.

The ReO(DMSA)$_2^-$ compound is of interest to coordination chemists because of the possibility of more than one isomer being produced in the reaction.

These isomers are produced due to the fact that in this reaction two chiral bidentate chelating DMSA ligands bind to the Re=O metal center. The ligand DMSA is meso-dimercaptosuccinic acid. The meso conformation is due to the configuration of the two chiral centers on the ligand.

SYN-ENDO

SYN-EXO

ANTI

The three ReO(DMSA)$_2^-$ isomers can be identified by proton NMR since the isomers each have their own set of peaks. The syn-endo and the syn-exo have singlets at ~ 4.13 ppm and 4.23 ppm respectively.

The anti isomer has two singlets at ~4.16 and 4.19 ppm. These peaks are due to the protons on the DMSA backbone.

The infrared spectra of both perrhenate and the product should be compared as KBr pellets or in a mull. The perrhenate ion has a characteristic Re-O stretch in the 800-1000 cm^{-1} region. The product has a strong Re=O stretch in the same region. Also obtain the IR spectrum of meso-dimercaptosuccinic acid. The S-H stretching frequency seen in the IR spectrum of meso-dimercaptosuccinic acid should not be present in the product.

Chemicals Needed:
(a) potassium perrhenate
(b) meso-dimercaptosuccinic acid
(c) stannous chloride dihydrate
(d) tetrabutyl ammonium bromide
(e) anion exchange resin
(f) cation exchange resin

Procedure
(1) In a 20 ml round-bottomed distillation flask (or, alternatively a 20 mL scintillation vial) weigh out
____________ grams (0.75mmol) of potassium perrhenate (KReO$_4$).

(2) Add to this vessel ___________ grams (1.5 mmol) of meso-dimercaptosuccinic acid.

(3) Add 10 ml of deionized water to the flask to make a suspension and place a magnetic stir bar in the reaction vessel to stir the reaction mixture while on a stirring hotplate.

(4) Next, add ___________ grams (0.75 mmol) of stannous chloride.

(5) Place the flask in a preheated 80-95° C water bath on a stirring hotplate after it has been tightly capped, and let it react for 30 minutes at this temperature with stirring. (try to keep solids down in the liquid)

(6) Remove the flask from the heating bath and allow it to cool.

(7) Filter the cooled solution.

(8) Determine the charge on the colored complex by taking out 250 ul of the filtered solution and placing on an anion exchange column, and 250 ul on a cation exchange column. Wash both loaded columns with 5 ml of water. The columns can be constructed by placing about 1 mL of the dry resin in a Pasteur pipette. Be sure to wash the resins thoroughly with distilled water prior to making the column bed.

(9) Take the rest of the filtered solution and add ________grams (1.50 mmol) of tetrabutyl ammonium bromide in 1 ml of water to the solution. Stir and filter off the precipitate.

(10) Let the precipitate dry until next lab period and obtain the IR spectrum in the 800-1000 cm^{-1} region.

(11) Obtain the ^1H NMR spectrum in CD_3OD, and also in D_6-DMSO.

QUESTIONS to be answered in the Discussion and report:
1. In the reaction, which metal is being oxidized and which is being reduced- Re or Sn?
2. What are the oxidation states of Re before and after the reaction?
3. What are the oxidation states of Sn before and after the reaction?
4. Why is the product colored, yet the perrhenate ion is colorless? (in general terms of d orbitals and electrons) .
5. What is the structure (how many bonds) and charge on the ReO core?
6. Draw the isomers of the metal complex (syn-endo, syn-exo, and anti).

References

(1) Bisunadan M. M., Blower P. J., Clarke S. E. M., Singh J., Went M. J., (1991) Appl. Radiat. Isot. Vol. 42, No.2, pp. 167-171.

(2) Hirano T., Tomiyoshi K., Zhang Y. J., Ishida T., Inoue T., Endo K., European Journal Of Nuclear Medicine, Vol. 21, No. 1, Jan. (1994) p.82-85

(3) Blower P. J., Singh J., Clarke S.E.M., J. Nucl. Med. 1991; 32:845-849

(4) Horiuchi K., Yomoda I., Ohta H., Endo K., Yokoyama A., Eur. J. Nucl. Med. (1991) 18:796-800

(5) Watkinson J. C., Lazarus C. R., Mistry R., Shaleen O.H., Maisey M.N., Clarke S., J. Nucl. Med. 30:174-180 (1989)

(6) Singh J., Powell A.K., Clarke S.E.M., Blower P.J., J. Chem. Soc., Chem. Commun., 1991, p1115

(7) Jurisson S., Berning D., Jia W., Ma D., "Coordination Compounds in Nuclear Medicine", Chem. Rev. 1993, 93, 1137-1156

(8) Chilton, H.M.; Witcofski, R.L. In "Pharmaceuticals in Medical Imaging"; Swanson, D.P., Chilton, H.M., Thrall, J.H., Eds.; Macmillan Publishing Co.: New York, 1990; p285

Chapter 1 Answers to Problems and Exercises

1. 2=linear or bent

4= square planar or tetrahedral

6= octahedral or trigonal planar or trigonal prismatic

2. (d) 3

3. (a) $[Cr(NH_3)_4Br_2]Br$

(b) $[Cr(NH_3)_4Br_2]Br \rightarrow [Cr(NH_3)_4Br_2]^+ + Br^-$

4. (a) six (b) octahedral (c) Fe^{3+}

5. (a) six (b) octahedral (c) Mn^{3+}

6. (a) six (b) octahedral (c) Co^{3+}

7. (a) six (b) octahedral (c) Co^{3+}

8.

Compound	Ionic nitrites	Isomers
(1) $[Co(NH_3)_6](NO_2)_3$	three	1
(2) $[Co(NH_3)_5NO_2](NO_2)_2$	two	1
(3) $[Co(NH_3)_4(NO_2)_2]NO_2$	one	2 (cis and trans)
(4) $[Co(NH_3)_3(NO_2)_3]$	0	2 (fac and mer)
(5) $Na[Co(NH_3)_2(NO_2)_4]$	0	2 (cis and trans)
(6) $Na_2[Co(NH_3)(NO_2)_5]$	0	1
(7) $Na_3[Co(NO_2)_6]$	0	1

9.

Compound	Ionic chlorides	Isomers
(1) $[Fe(H_2O)_6]Cl_2$	two	1
(2) $[Fe(H_2O)_5Cl]Cl$	one	2 (cis and trans)
(3) $[Fe(H_2O)_4Cl_2]$	0	2 (cis and trans)
(4) $NH_4[Fe(H_2O)_3Cl_3]$	0	2 (fac and mer)
(5) $(NH_4)_2[Fe(H_2O)_2Cl_4]$	0	2 (cis and trans)
(6) $(NH_4)_3[Fe(H_2O)Cl_5]$	0	1
(7) $(NH_4)_4[FeCl_6]$	0	1

10.
Compound
(1) $[Pt(NH_3)_4](SCN)_2$
(2) $[Pt(NH_3)_3(SCN)]SCN$
(3) $[Pt(NH_3)_2(SCN)_2]$
(4) $NH_4[Pt(NH_3)(SCN)_3]$
(5) $(NH_4)_2[Pt(SCN)_4]$

11. (a)
(1) $[Pd(PØ_3)_4]Cl_2$
(2) $[Pd(PØ_3)_3Cl]Cl$
(3) $[Pd(PØ_3)_2Cl_2]$
(4) $NH_4[Pd(PØ_3)Cl_3]$
(5) $(NH_4)_2[PdCl_4]$
(b) Compounds (1) and (5) would have the greatest conductivity in solution because (1) is a 1:2 electrolyte, and (5) is a 2;1 electrolyte. Each produce three equivalents of ions, the highest number for this series of compounds.
(c) $[Pd(PØ_3)_2Cl_2]$, **two isomers, cis and trans**

12. Three ions; 2 Cl^- ions and $[Co(NH_3)_5Cl]^{2+}$

13. Five ions; 4 K^+ ions and $[FeCl_6]^{4-}$

14. Three moles of bromide ion

15. $1.96g \times 1mol/267.28g = 7.33 \times 10^{-3}$ **mol** $[Co(NH_3)_6]Cl_3$
7.33×10^{-3} **mol** $[Co(NH_3)_6]Cl_3 \times 3mol\ Cl^-/mol$
$[Co(NH_3)_6]Cl_3 = 0.0220$ **mol** Cl^-
0.0220 mol Cl^- **x 143.32 g AgCl/1mol AgCl = <u>3.15 g AgCl</u>**

16. $2.54g \times 1mol/949.07g = 2.676 \times 10^{-3}$ **mol** $Ba_3[FeCl_6]_2$
2.676×10^{-3} **mol** $Ba_3[FeCl_6]_2 \times 3mol\ Ba^{2+}/mol\ Ba_3[FeCl_6]_2 =$
8.03×10^{-3} **mol** Ba^{2+}
8.03×10^{-3} **mol** Ba^{2+} **x 233.39 g** $BaSO_4$**/mol = <u>1.87 g $BaSO_4$</u>**

17. 5g x 1mol/366.63g = 0.0136 mol [Co(NH$_3$)$_4$Br$_2$]Br
0.0136 mol Br$^-$/0.100 L = <u>0.136 M Br$^-$</u>

18. One [Rh(NH$_3$)$_5$Cl]$^{2+}$ ion and two Cl$^-$ ions

19. Three NH$_4^+$ ions and a [FeCl$_6$]$^{3-}$ ion

20. Two isomers for square planar, and one isomer for tetrahedral:

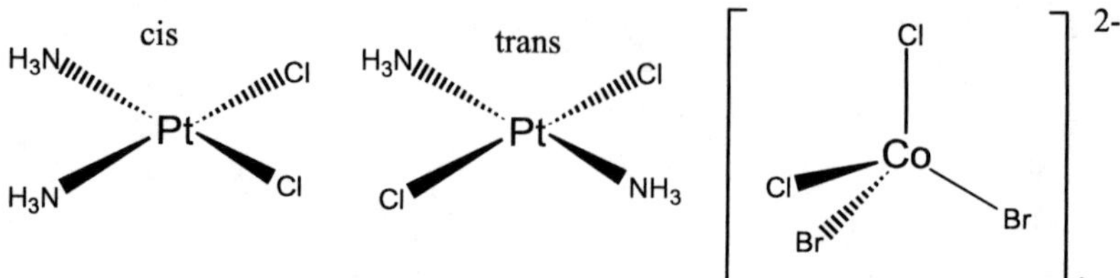

21.
(a) Os = [Xe]6s^{2}5d^{6}4f^{14}
(b) Co = [Ar]4s^{2}3d^7
(c) Ni = [Ar]4s^{2}3d^8
(d) Ru = [Kr]5s^{1}4d^7
(e) Cu = [Ar]4s^{1}3d^{10}
(f) Re = [Kr]5s^{2}4d^5

22.
(a) Zr = +4
(b) Ta = +5
(c) Mn = +7
(d) Nb = +5
(e) Tc = +7
(f) Y = +3

23.
(a) [Ar] 3d^6 = Fe^{2+}, Co^{3+}
(b) [Ar] 3d^5 = Mn^{2+}, Fe^{3+}
(c) [Ar] 3d^{10} = Cu$^+$, Zn^{2+}
(d) [Ar] 3d^8 = Ni^{2+}, Co$^+$

24.(a) $Pt^{2+} = 5d^8$ (b) $Ni^{2+} = 3d^8$ (c) $Co^{3+} = 3d^3$ (d) $Ir^{3+} = 5d^3$
(e) $Ti^{4+} = [Ar]$ (f) $Zr^{4+} = [Kr]$ (g) $Cu^{2+} = 3d^9$ (h) $Ag^+ = 4d^9$
(i) $Cr^{3+} = 3d^3$ (j) $Mo^{3+} = 4d^3$ (k) $Mn^{3+} = 3d^4$ (l) $Tc^{3+} = 4d^4$

25. six

26. Ir^{3+}

27. (c) the lanthanide contraction.

28.

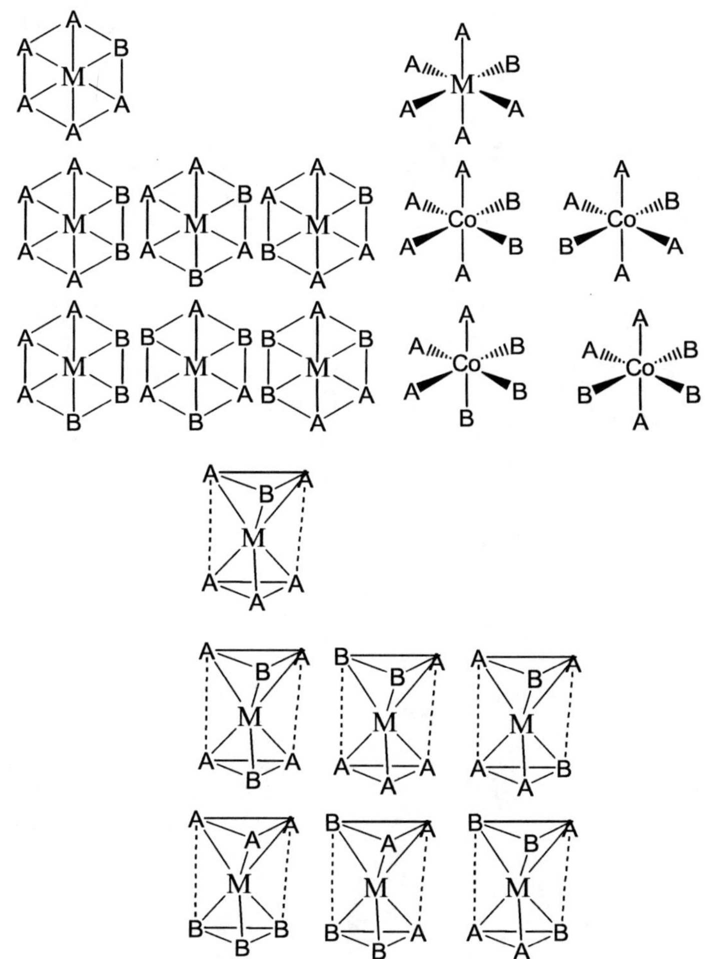

Advanced
29. (a)

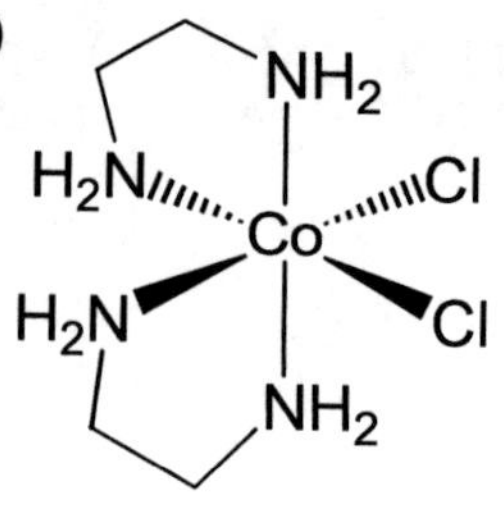
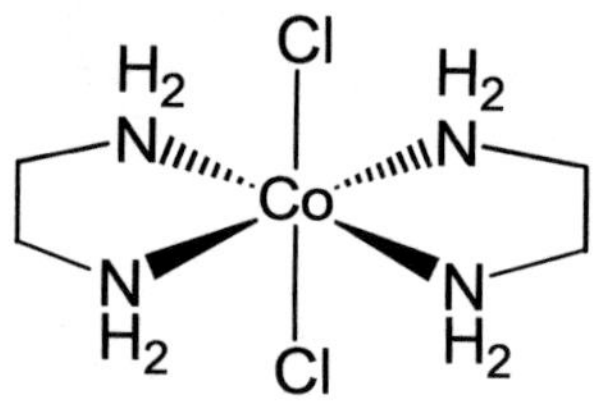

(b)

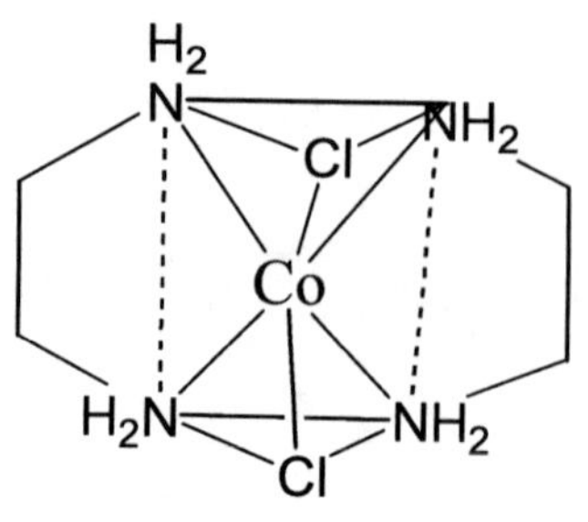
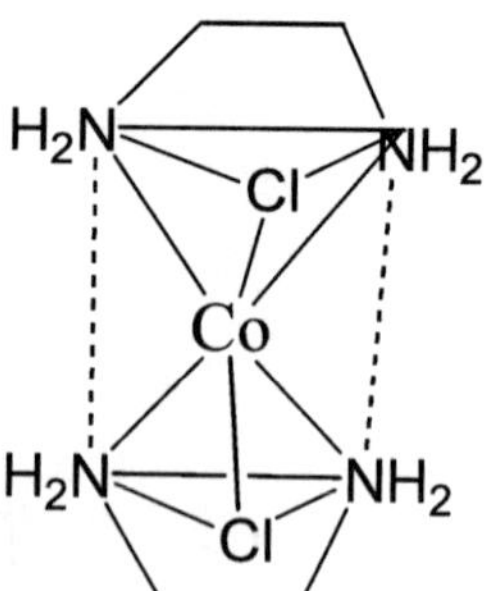

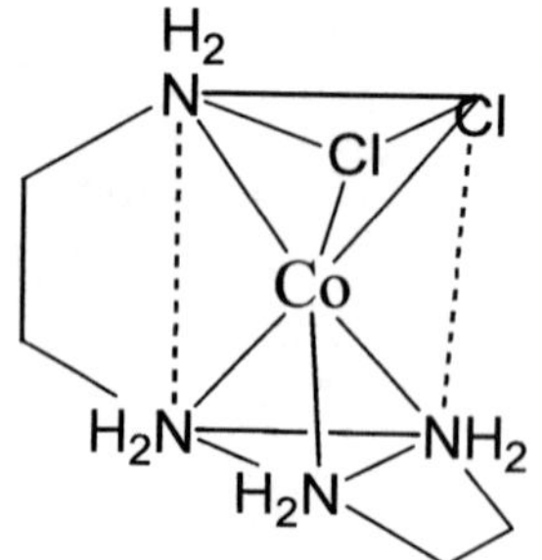
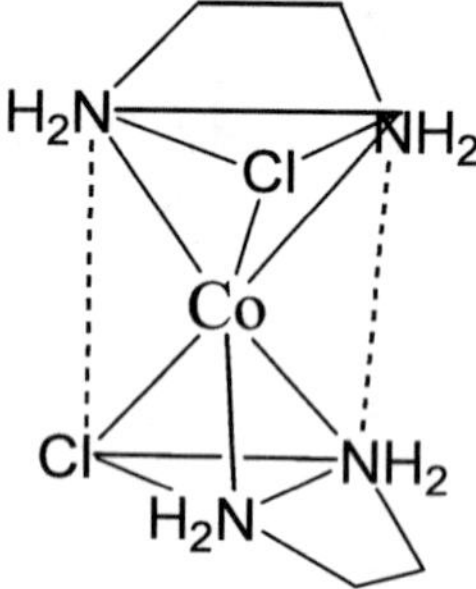

(c)

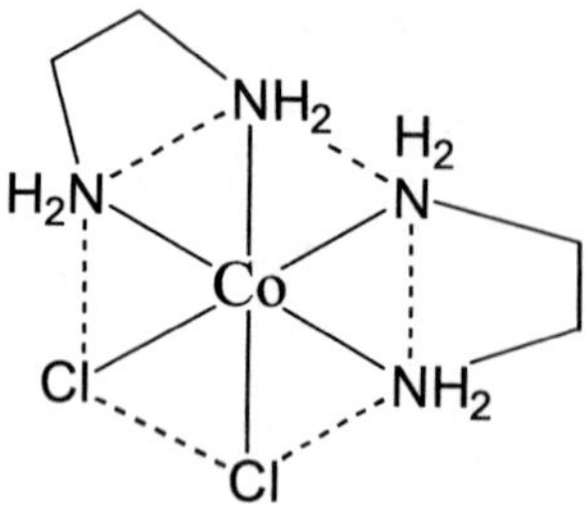

Chapter 2 Answers to Problems and Exercises

1.

(A) O_3 (C) CO_3^{2-} (D) NO_2^{-}

2. three

3. One

4. $C = -2$ $S = +2$ $N = -1$
$$[::C=S=N::]^{-}$$

5. $N = -1$ $C = 0$ $S = 0$
$$[::N=C=S::]^{-}$$

6. (A) NO

7. Three

8. Al^{3+}, BF_3

9. Tetrahedral

10. Linear

11. Trigonal Planar

12. Slightly less than 109.5°

13. Trigonal Planar

14. 90° and 120°

15. (A) COS

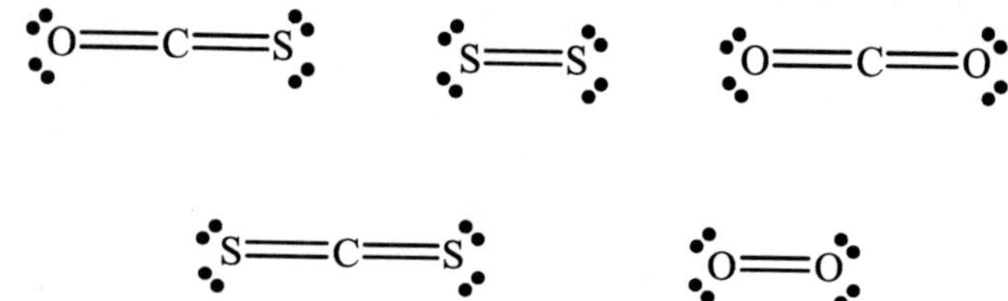

16. (A) CH₂F₂

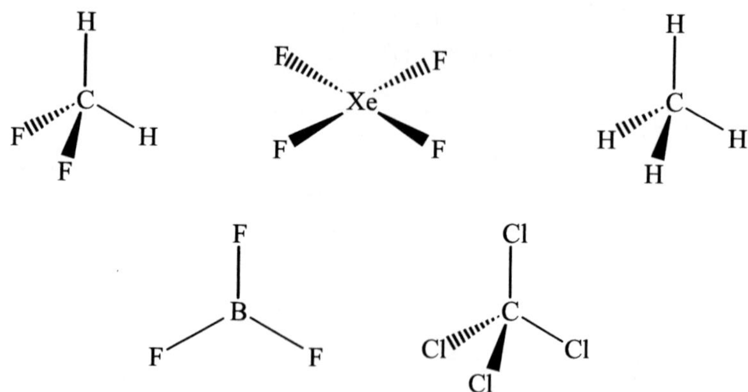

17. (B) XeF₂

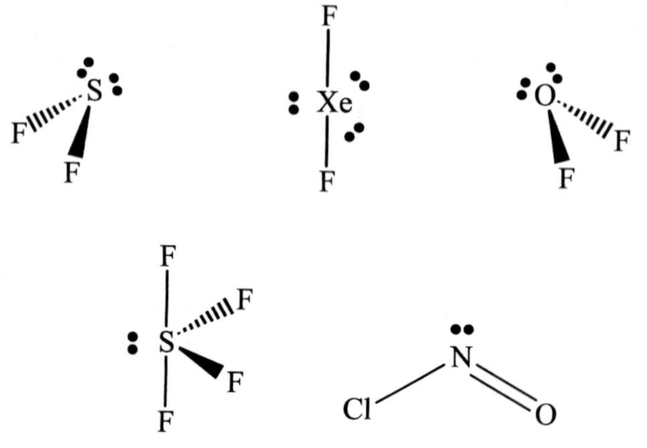

216

18.

(a) $Fe^{2+} = 3d^6$, $Fe^{3+} = 3d^5$, $\underline{Sc^{3+} = [Ar]}$, $Co^{3+} = 3d^6$

(b) $Tl^+ = 6s^2 5d^{10}$, $\underline{Te^{2-} = [Xe]}$, $Cr^{3+} = 3d^3$

(c) $Pu^{4+} = 5f^4$, $\underline{Ce^{4+} = [Xe]}$, $Tl^{3+} = 5d^{10}$

(d) $\underline{Ba^{2+} = [Xe]}$, $Pt^{2+} = 5d^8$, $Mn^{2+} = 3d^5$

19.

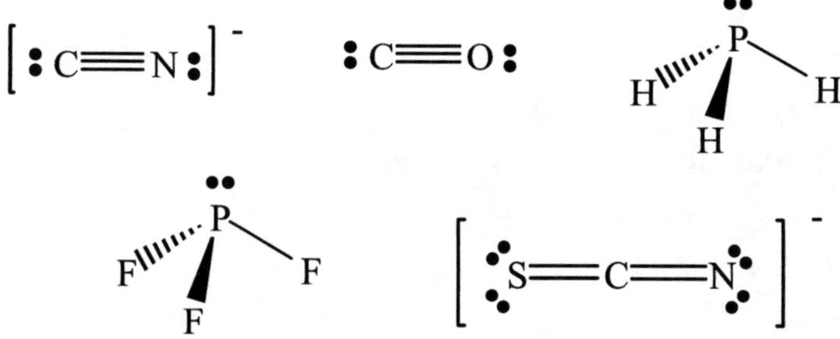

20. Fluorine (a second row element with no possible d-orbital involvement) cannot have an expanded valence shell that holds more than eight electrons.

Key for Chapter 2 Practice Quiz

1. (e) all contain polar bonds
2. (e) none contain polar bonds
3. (e) none have a dipole moment
4. (e) NH_3
5. Tetrahedral = (b) H_2O
6. Trigonal Bipyramidal = (c) PCl_5
7. Octahedral = (d) XeF_4
8. Trigonal Planar = (e) NO_2^-
9. Linear = (a) CO_2
10. (a) true
11. Tetrahedral = (a) CH_4
12. Trigonal Bipyramidal = (c) PCl_5
13. Square Planar = (d) XeF_4
14. Bent = (b) H_2O
15. See-Saw = (e) SF_4^-
16. (c) tetrahedral
17. (a) pyramidal
18. (a) The ammonia molecule has tetrahedral molecular geometry.
19. (b) trigonal planar
20. (c) tetrahedral
21. (a) a single covalent bond.
22. (b) two
23. (a) one
24. (b) O_3
25. (b) two
26. (d) F-I
27. (e) $SbCl_3$
28. (d) O_2

Chapter 3 Answers to Problems and Exercises

1.

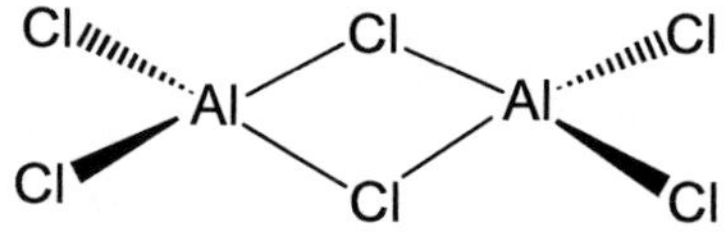

$$Cl_2Al(\mu\text{-}Cl)_2AlCl_2$$

2.
(A) ethylenediamine = bidentate
(B) aqua = monodentate
(C) cyano = monodentate
(D) carbonyl = monodentate
(E) ammine = monodentate

3. The metal is a <u>Lewis Acid</u> and the ligands are Lewis Bases.

4. (a) chloride (d) CH_3^- (e) iodide

5. a Lewis acid

6.
   (a) I-
   (b) S^{2-}
   (c) AsH_3
   (d) dppe
   (e) thiocyano
   (f) dtc (dithiocarbamate)
   (g) pamp

7. (a) Cu^{2+} (b) Ni^{2+} (c) Cr^{3+}

8. (a) diphos (b) EDTA
   (c) acac (d) diop
   (e) porphyrin (f) dmpe
   (g) cyclam

9..

    (a) ammonia = classical
    (b) water = classical
    (c) chloride = classical
    (d) EDTA = classical
    (e) en = classical
    (f) diphos = nonclassical
    (g) acac = classical
    (h) DuPhos = nonclassical
    (i) porphyrin = classical
    (j) dipamp = nonclassical
    (k) carbon monoxide = nonclassical
    (l) methylamine = classical

10.

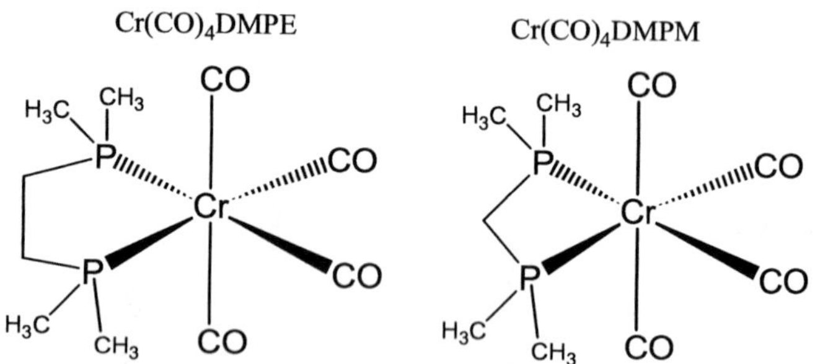

The DMPE ligand forms a six-membered chelate ring, but the DMPM ligand forms a more strained five-membered ring.

11.

(a) $[Pt(NH_3)_4Cl_2]SO_4$
dichlorotetraammine platinum(IV) sulfate

(b) $K_3[Mo(CN)_6F_2]$
potassium difluorohexacyanomolybdenate(V)

(c) $K[Co(EDTA)]$
potassium ethylenediamminetetracetatocobaltate(III)

(d) $[Co(NH_3)_3(NO_2)_3$
trinitrotriammine cobalt(III)

12.
(a) $[Pt(NH_3)_6]Cl_4$
hexaammineplatinum(IV) chloride

(b) $[Ni(acac)(P(C_6H_5)_3)_4]NO_3$
acetylacetonatotetrakistriphenylphosphine nickel(II) nitrate

(c) $(NH_4)_4[Fe(ox)_3]$
ammonium trisoxalato ferrate(II)

(d) $W(CO)_3(NO)_2$
dinitrotricarbonyl tungsten(II)

13.
(a) $[Pt\{P(C_6H_5)_3\}_4](CH_3COO)_4$
tetrakistriphenylphosphine platinum(IV) acetate

(b) $Ca_3[Ag(S_2O_3)_2]_2$
calcium bisthiosulfato argentate(I)

(c) $Ru(As(C_6H_5)_3)_3Br_2$
dibromotristriphenylarsine ruthenium(II)

14.
(a) $[Fe(en)_3][IrCl_6]$
trisethylenediammine iron(III) hexachloroiridate(III)

(b) $[Ag(NH_3)(CH_3NH_2)]_2[PtCl_2(ONO)_2]$
amminemethylaminesilver(I) dichlorodinitrito platinum(II)

(c) $[VCl_2(en)_2]_4[Fe(CN)_6]$
dichlorobisethylenediammine vanadium(III)
hexacyanoferrate(II)

15.

(a) Pentaammine(dinitrogen)ruthenium(II) chloride
$[Ru(NH_3)_5N_2]Cl_2$

(b) Aquabis(ethylenediamine)thiocyanatocobalt(III) nitrate
$[Co(H_2O)(en)_2(SCN)](NO_3)_2$

(c) Sodium hexaisocyanochromate(III)
$Na_3[Cr(NC)_6$

16.

(a) Bis(methylamine)silver(I) acetate
$[Ag(CH_3NH_2)_2](CH_3COO)$

(b) Barium dibromodioxalatocobaltate(III)
$Ba_3[Co(ox)_2Br_2]_2$

(c) Carbonyltris(triphenylphosphine)nickel(0)
$[Ni(CO)(P\emptyset_3)_3]$

17.

(a) Tetrakis(pyridine)bis(triphenylarsine)cobalt(III) chloride
$[Co(pyr)_4(As\emptyset_3)_2]Cl_3$

(b) Ammonium dicarbonylnitrosylcobaltate(-I).
$NH_4\,[Co(CO)_2(NO)]$

(c) Potassium octacyanomolybdenate(V)
$K_3[Mo(CN)_8]$

(d) Diamminedichloroplatinum(II)
$[Pt(NH_3)_2Cl_3]$

18.

(a) $[Co(NH_3)_6][CuCl_5]$
(b) $[Pt(pyr)_4][PtCl_4]$
(c) $[Pd(NH_3)_2(P\emptyset_3)_2]$
(d) $[Au(ox)_2]^-$

19. **More than one valid structure may be drawn for these compounds; they are isomers. Here is one:**

(a)

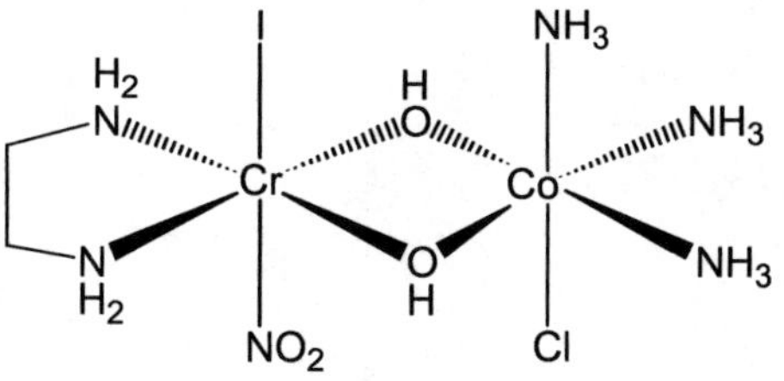

(b)

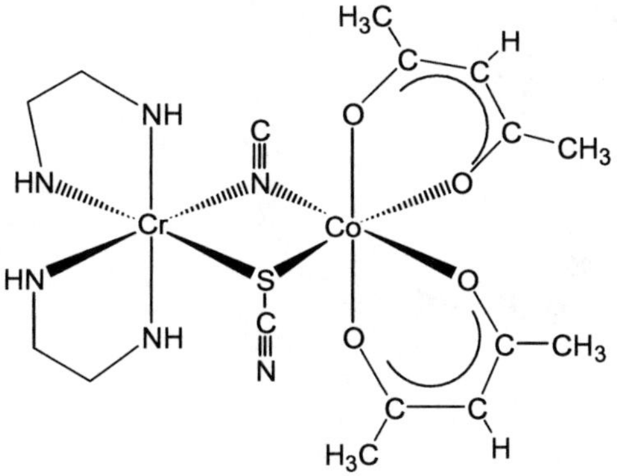

20.

(a)

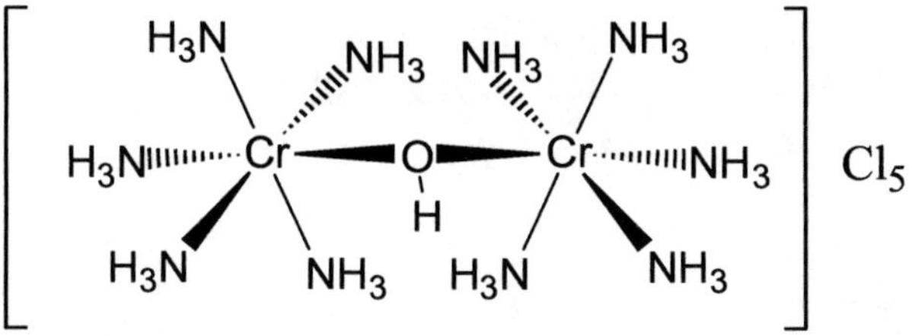

(b)

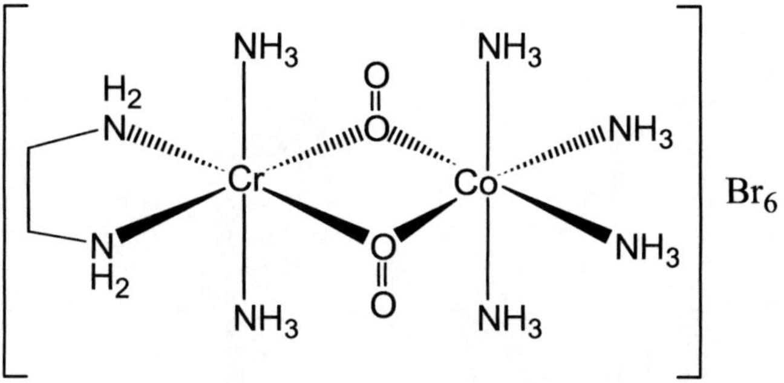

21.

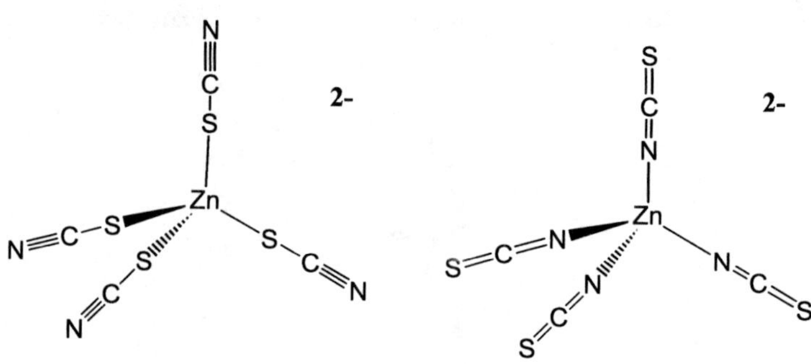

22.

   (a) dmpm
   (b) dppm
   (c) dppe
   (d) dipamp
   (e) triphenyl phosphite
   (f) trien

23.

   (a) $Co_4(CO)_{12}$ = Homobimetallic
   (b) $Co_3Rh(CO)_{12}$ = Heterobimetallic
   (c) $Mn_2(CO)_{10}$ = Homobimetallic
   (d) $Fe_2Ru(CO)_{12}$ = Heterobimetallic

24.

(a) $\log \beta_6 = 2.79 + 2.26 + 1.69 + 1.25 + 0.74 + 0.03 = 8.76$
(b) $\beta_6 = 5.75 \times 10^8$

Chapter 4 Answers to Problems and Exercises

1.

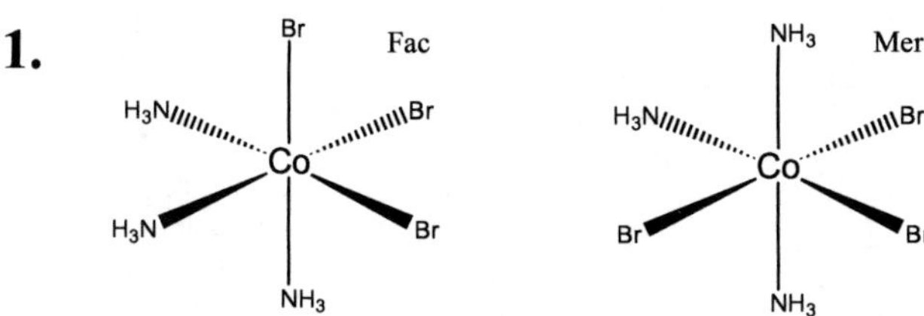

2.

3.

4.

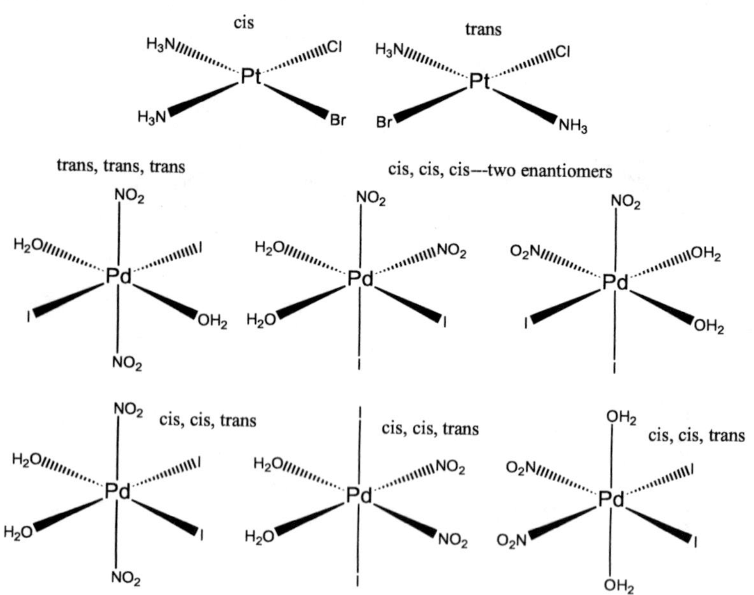

5. (a) and (b)

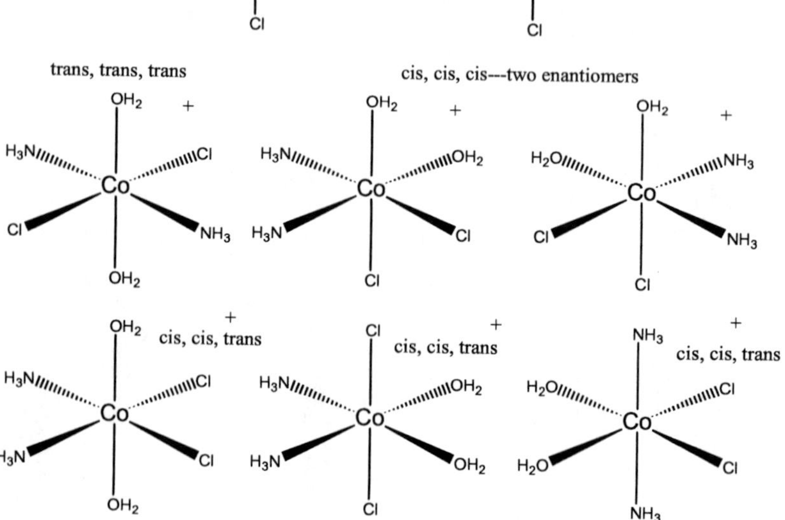

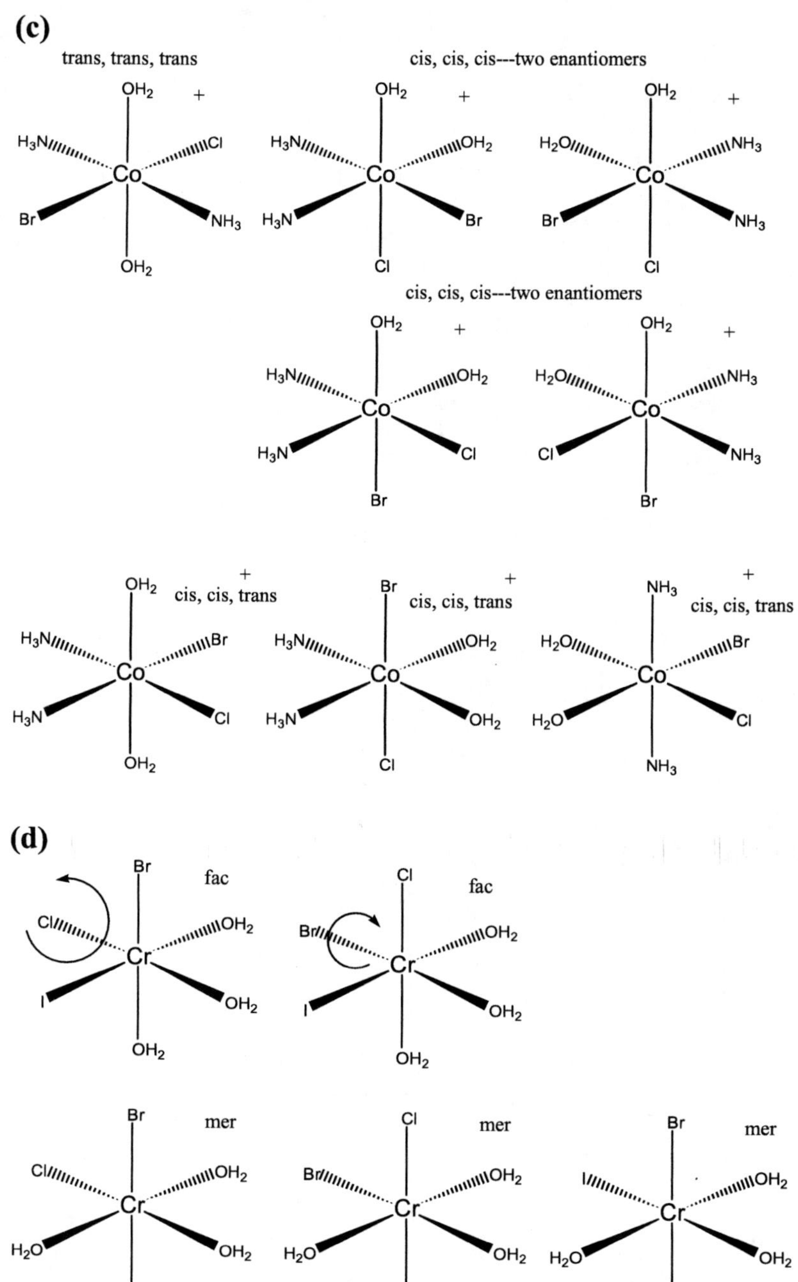

227

(e)

cis
trans
2+
2+

(f)

fac
mer

6.

Mer-cis
Mer-trans
Fac

7.

Geometrical isomers
Optical isomers
2+
2+
2+

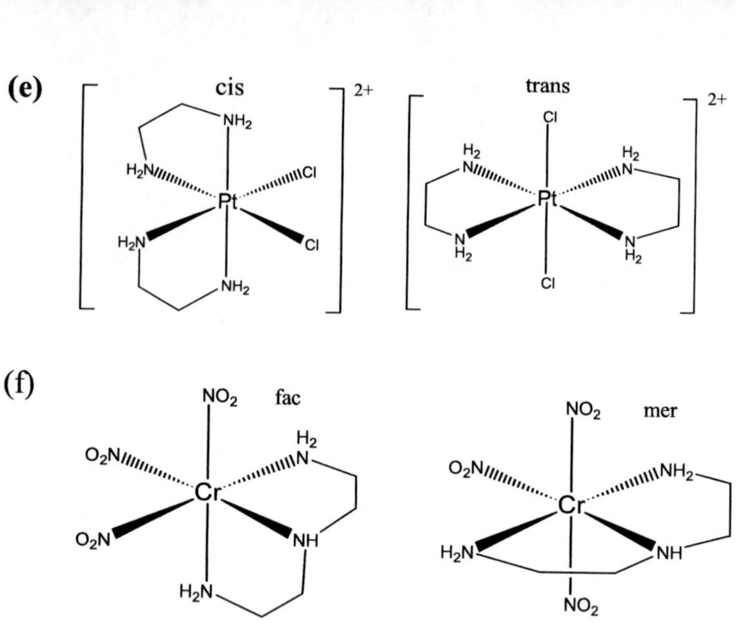

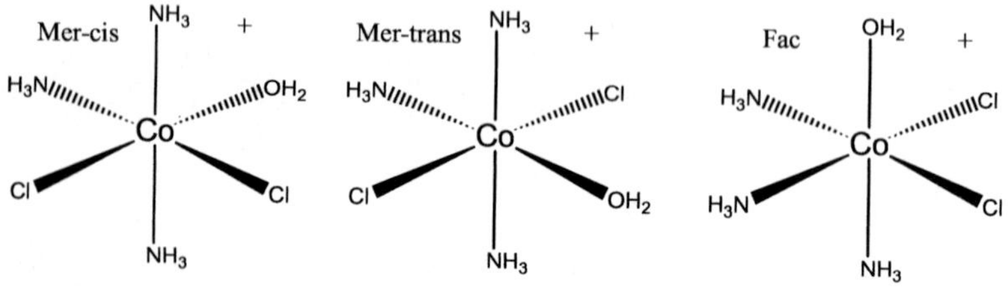

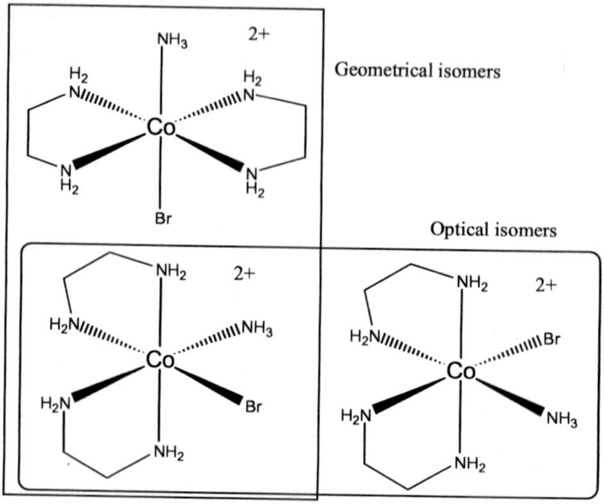

8.

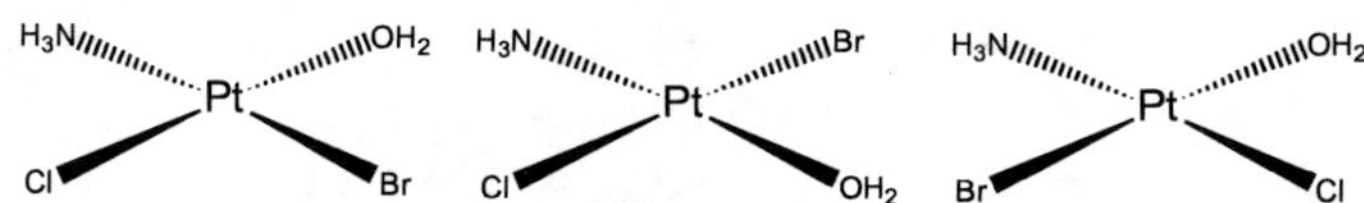

9. Chiral=nonsuperimposable mirror images

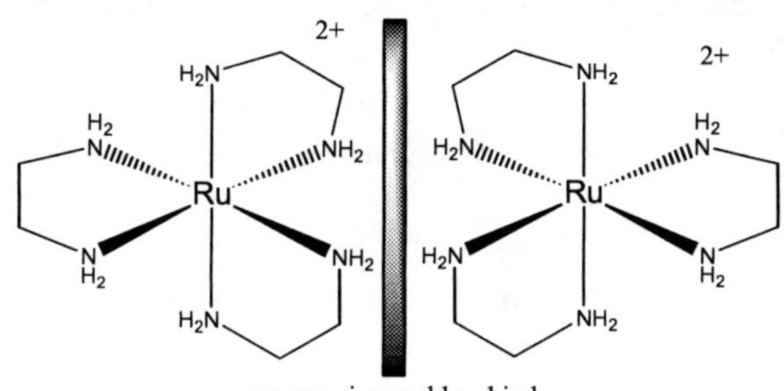

nonsuperimposable=chiral

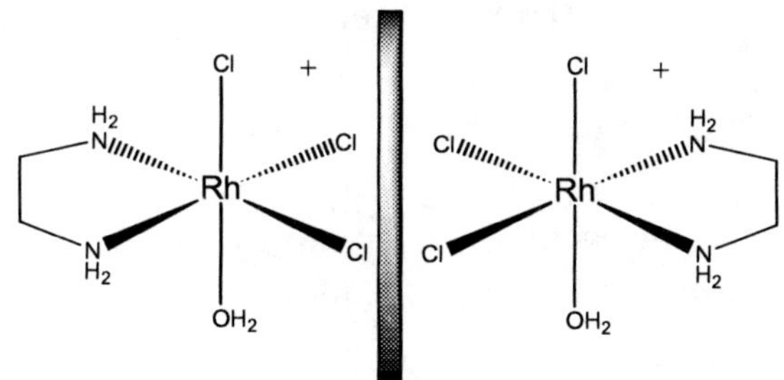

superimposable=nonchiral

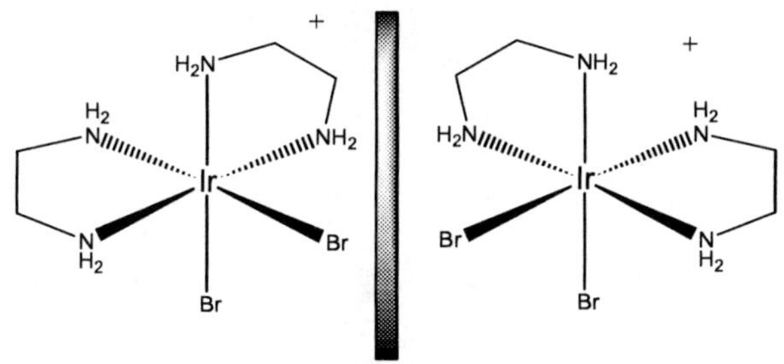

nonsuperimposable=chiral

10.

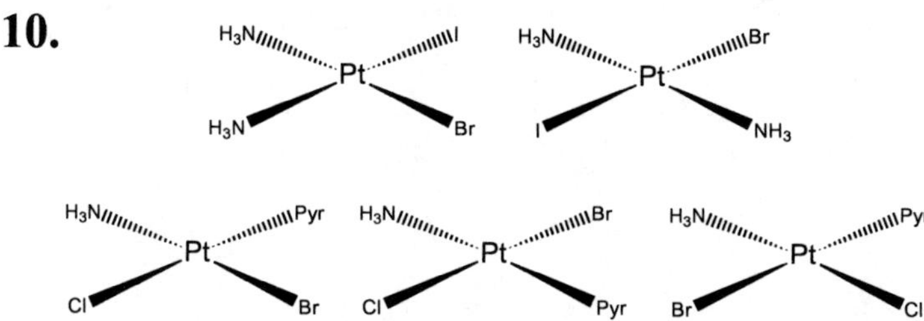

11.

Pt 38.8% or 38.8g x 1mol/195.08g = 0.198 mol Pt
Cl 14.1%   14.1g x 1mol/35.45g = 0.397 mol Cl
C 28.7%   28.7g x 1mol/12.04g =   2.39 mol C
P 12.4%   12.4g x 1mol/195.08g = 0.400 mol P
H 6.02%   6.02g x 1mol/195.08g = 5.95 mol H
   100% → 100g

thus, dividing by smallest, the empirical formula is:
Pt   0.198/0.198 = 1 Pt atom
Cl   0.397/0.198 = 2 Cl atoms
C   2.390/0.198 = 12 C atoms $Pt(P(ethyl)_3)_2Cl_2$
P   0.400/0.198 = 2 P atoms
H   5.95/0.198 = 30 H atoms
These must be cis and trans isomers of square planar Pt^{2+}

12. (c) geometric isomers

13. (b) nonsuperimposable mirror images with identical chemical formulae and the same chemical reactivities.

14. (c) linkage isomers.

15. only one

16.

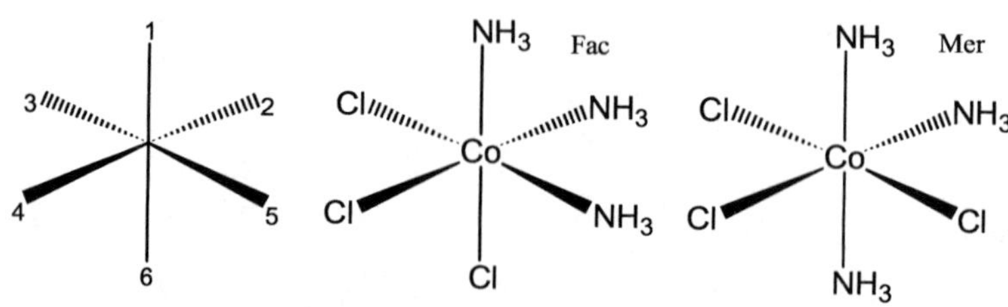

17. trans

18.

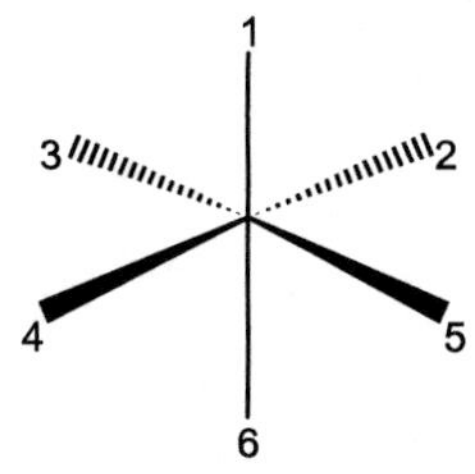

cis, cis, cis---two enantiomers

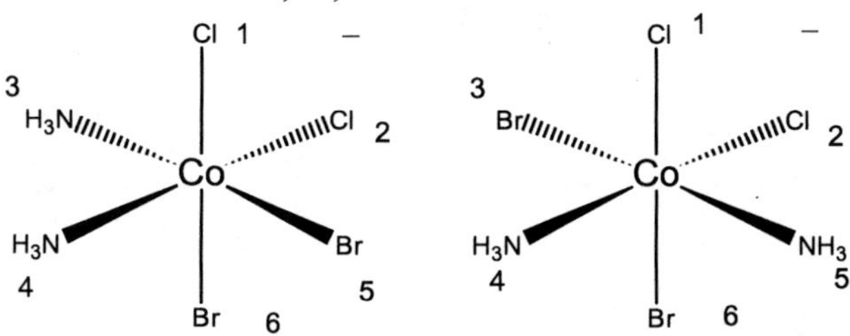

[Co(5,6-dibromo)(1,2-dichloro)(3,4-diammine)]

[Co(3,6-dibromo)(1,2-dichloro)(4,5-diammine)]

19.

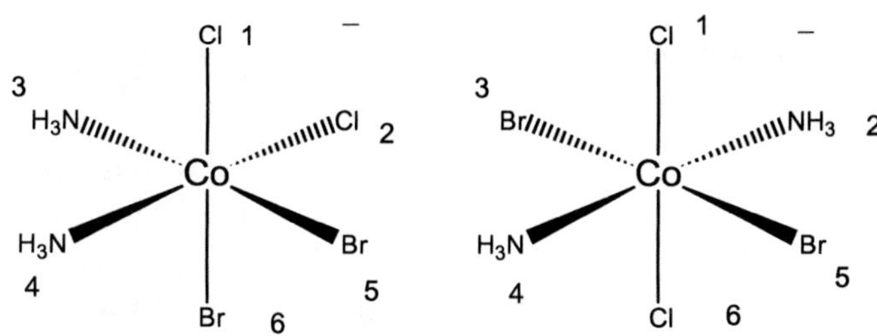

[Co(5,6-dibromo)(1,2-dichloro)(3,4-diammine)]

[Co(3,5-dibromo)(1,6-dichloro)(2,4-diammine)]

Chapter 5 Answers to Problems and Exercises

1. (d) colorless

2. (e) Cu+ and Zn^{2+}

3. (c) both d^6

4. None do   (a) d^8 (b) d^8 (c) d^6 (d) d^7 (e) d^6

5. $[Cr(CN)_6]^{4-}$ and $[Cr(OH_2)_6]^{2+}$

2 unpaired electrons   4 unpaired electrons

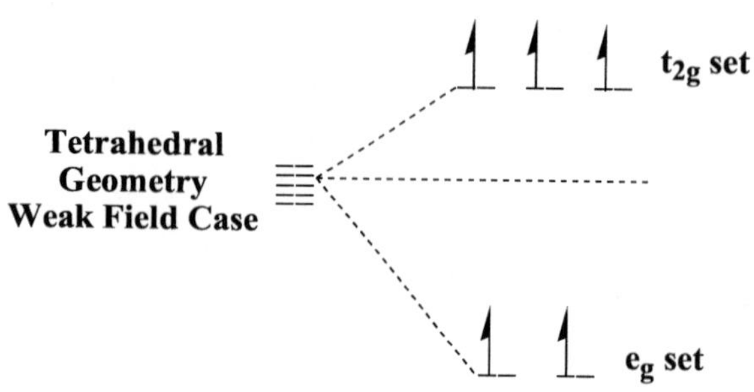

6. $[MnCl_4]^{2-}$ has a Mn^{2+} metal center. The Mn atom has a d^5 electron configuration. Since Cl⁻ is considered to be a weak field ligand and the tetrahedral geometry is always the weak field case, then the compound would have five unpaired electrons.

7. Lowest d-orbital splitting energy to the highest.
$[CoCl_6]^{3-} < [Co(ox)_3]^{3-} < [Co(OH_2)_6]^{3+} < [Co(NH_3)_6]^{3+} < [Co(CN)_6]^{3-}$

8. Lowest d-orbital splitting energy to the highest.
$[FeCl_4]^{2-} < [FeCl_6]^{4-} < [Fe(OH_2)_6]^{2+} < [Fe(NH_3)_6]^{2+} < [Fe(CN)_6]^{4-}$

9. (d) Co(II)

10. (a) the complex is high-spin.

11.

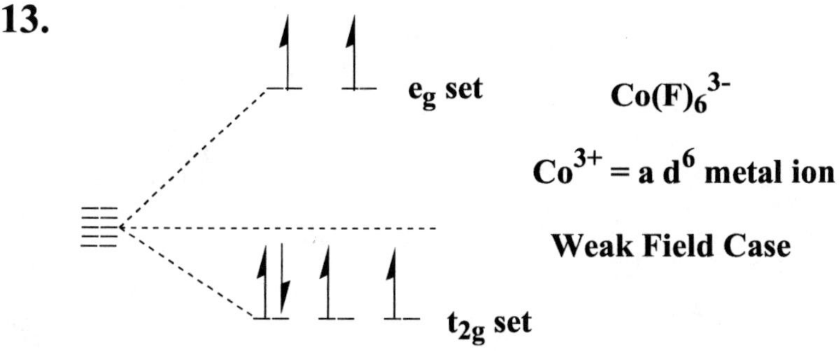

12. An octahedral complex is high-spin when the value of Δ_o is **less** than the energy required to **pair** the electrons.

13.

$[CoF_6]^{3-}$ $\mu_{Obs} = 5.3$ This complex is high spin.

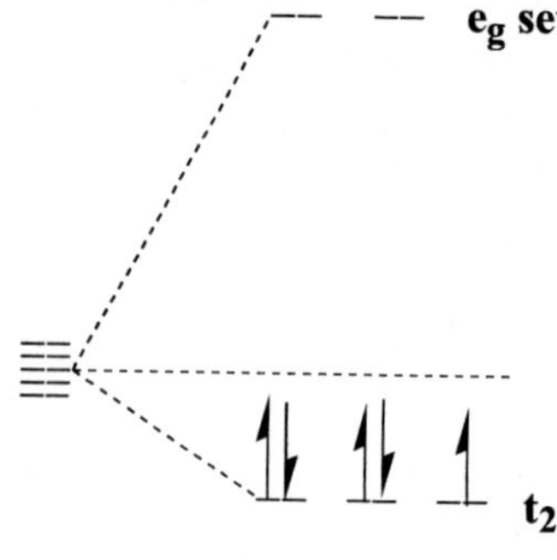

$[Fe(CN)_6]^{3-}$ $\mu_{Obs} = 2.3$ This complex is low spin.

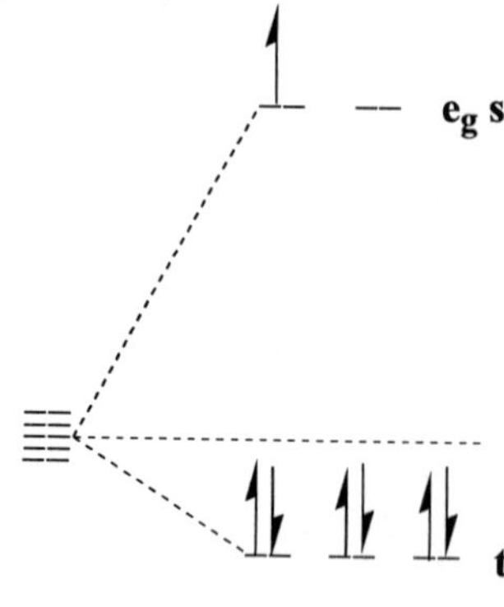

$[Co(NO_2)_6]^{4-}$ $\mu_{Obs} = 1.8$ This complex is low spin.

14. $[Cr(H_2O)_6]^{3+}$

15.

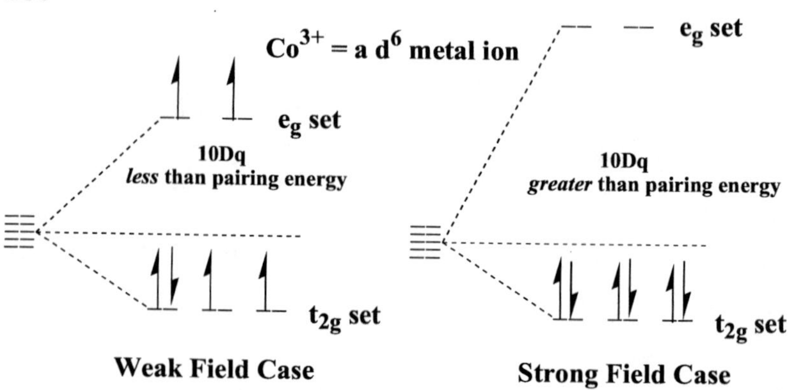

16.

$[Co(NH_3)_6]^{3+}$ appears orange-yellow   absorbs blue
$[Co(NH_3)_5Cl]^{2+}$ appears purple   absorbs yellow
$[Co(NH_3)_5(NCS)]^{2+}$ appears orange   absorbs blue-green
$[Co(NH_3)_5(H_2O)]^{3+}$ appears red   absorbs green

Using the wavelengths of the colors absorbed as an inverse relationship to the energy of Δ_o, then the compounds would have a ranking of from low energy to high energy Δ_o due to the colors absorbed. These colors are in the order: yellow<green<blue-green<blue. Thus the rankings are:

$[Co(NH_3)_5Cl]^{2+}$ absorbs yellow <

$[Co(NH_3)_5(H_2O)]^{3+}$ absorbs green<

$[Co(NH_3)_5(NCS)]^{2+}$ absorbs blue-green<

$[Co(NH_3)_6]^{3+}$ absorbs blue

Therefore the ligands field strength would be ranked as: Cl- < H_2O < NCS < NH_3 . This matches the order expected in the spectrochemical series.

17. $[Cr(Br)_6]^{3-} < [Cr(H_2O)_6]^{3+} < [Cr(NH_3)_6]^{3+}$

18.

The spectrochemical series lists the field strength of the ligands in the order: $Cl^- < en < CN^-$, so the metal complexes, which all have Cr(III), have a decreasing Δ_o in the order:

$[Cr(CN)_6]^{3-} > [Cr(en)_3]^{3+} > [Cr(Cl)_6]^{3-}$. The energy of the visible light absorbed would follow the same order.

19.   (b) Cl-

20. Jahn-Teller distortion may occur because of the degeneracy in the e_g set.

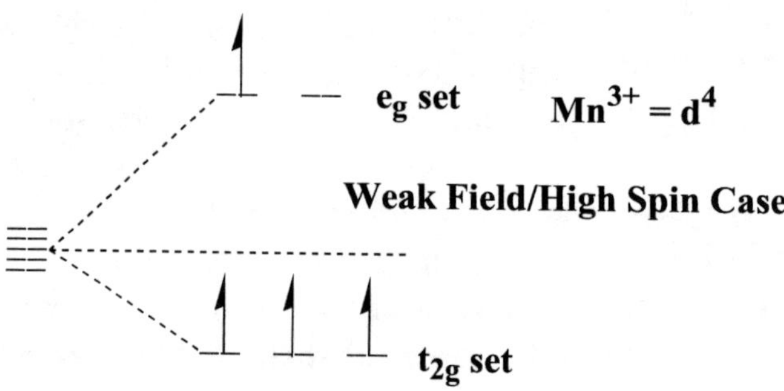

21. Jahn-Teller distortion could occur because of the degeneracy in the e_g set.

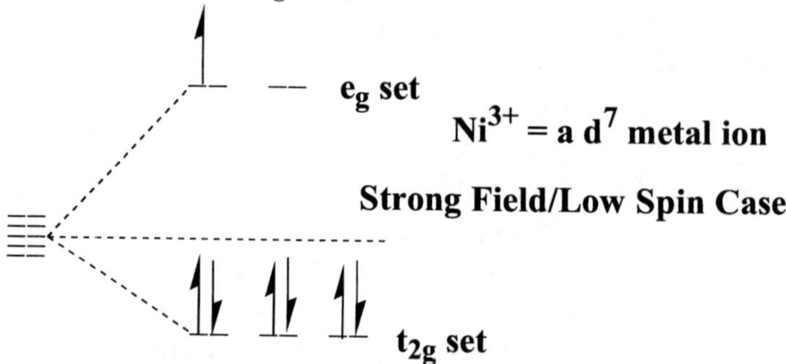

22. Cu^+ compounds are, by necessity, d^{10} metal centers, which cannot have d-d electron transitions since the d-subshell is filled. On the other hand Cu^{2+} compounds have a d^9 metal center, and d-d transitions in the visible range may occur.

23. Least amount of CFSE.

24. Greatest amount of CFSE.

KEY to Practice Quiz Chapter 5

1.

$[CoCl_6]^{3-}$ = **(III)** and it has **six** d electrons

$[Cr(CN)_6]^{4-}$ = **(II)** and it has **four** d electrons

$[CuCl_4]^{2-}$ = **(II)** and it has **nine** d electrons

$[Cr(OH_2)_6]^{2+}$ = **(II)** and it has **four** d electrons

2.

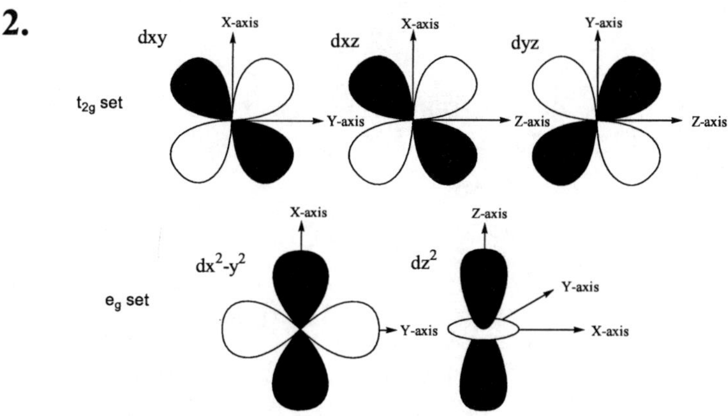

3. $[W(CN)_6]^{4-}$ **2 unpaired electrons,**
$[W(OH_2)_6]^{2+}$ **= 4 unpaired electrons.**

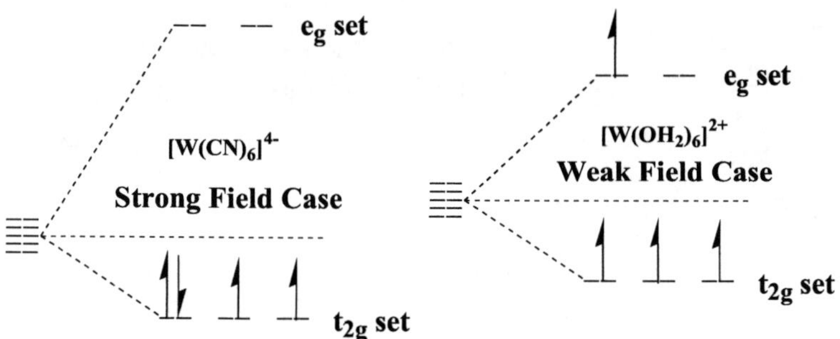

4. (A) 1

5. (C) $[RhCl_6]^{3-}$

6. (B) $[Os(CN)_6]^{4-}$

7. d^1 **through** d^3 **and then** d^8 **through** d^{10}

8.
(A) the complex is high-spin.
(D) the complex is paramagnetic
(F) Δ_o is smaller than the pairing energy.

9. (E) 4

10. FALSE

11. $[Tc(CN)_6]^{3-}$ has 2 unpaired electrons,
$[TcCl_6]^{4-}$ has 3 unpaired electrons.

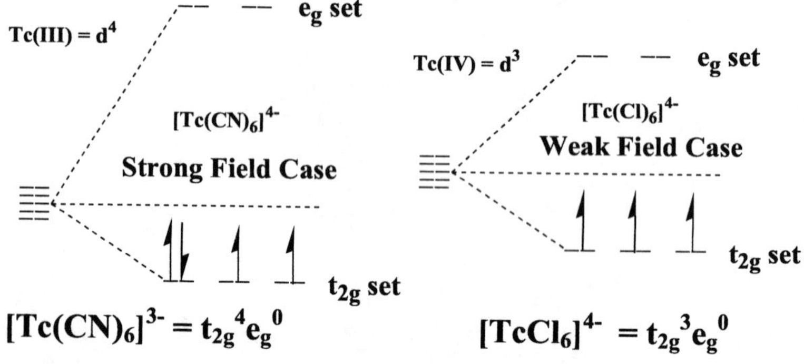

BONUS question: Tetrahedral geometry would be paramagnetic, only square planar is diamagnetic.

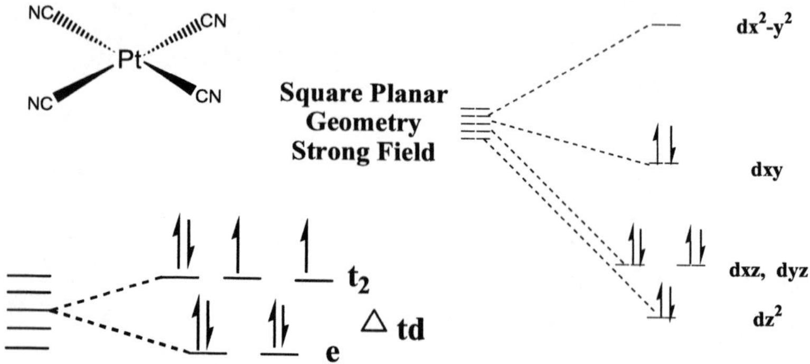

Chapter 6 Answers to Problems and Exercises

1.

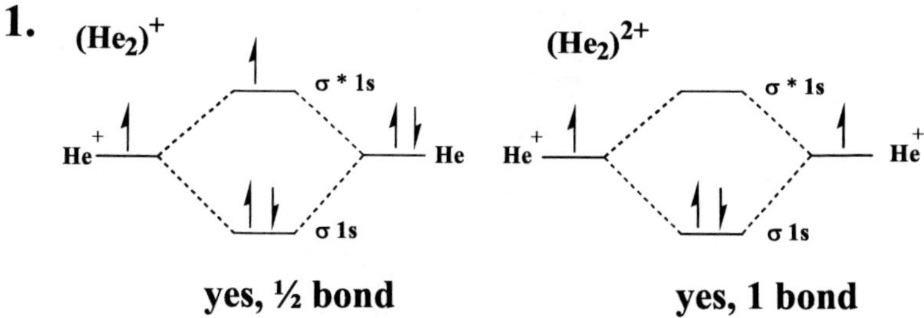

2.

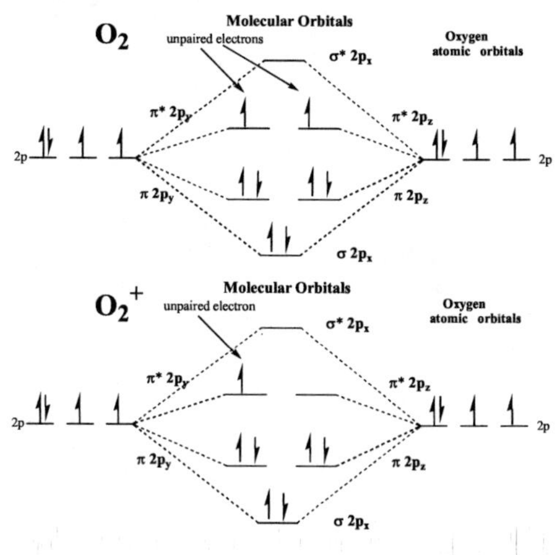

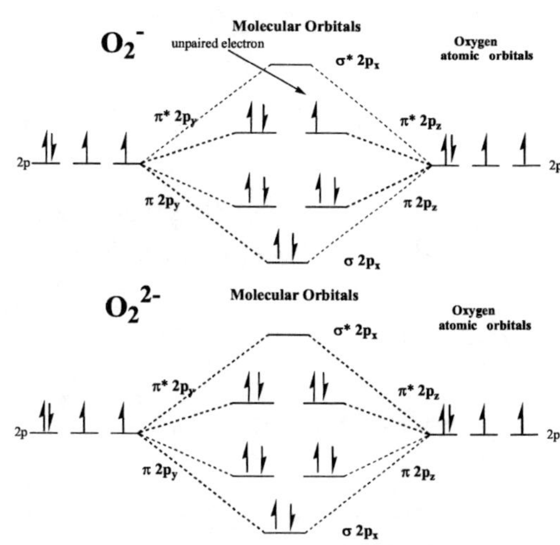

3.

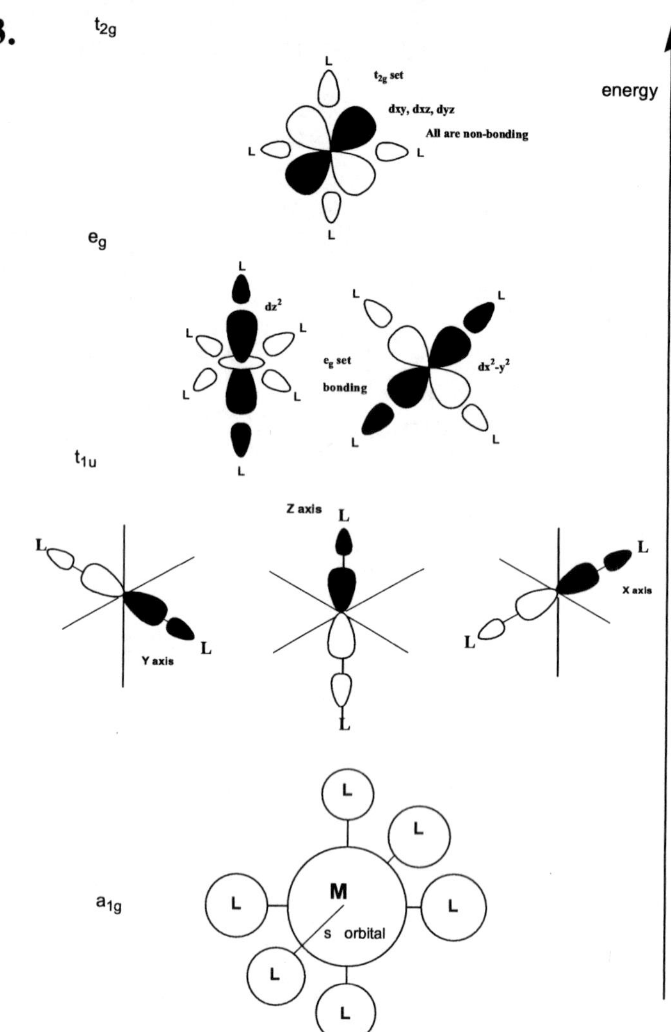

4. The a$_{1g}$, the t$_{1u}$, and the e$_g$ set are all bonding since there are effective overlaps, but the t$_{2g}$ set is nonbonding.

5. Scheme shown on page 159.

6. Scheme shown on page 160.

7. Scheme shown on page 161.

8.

Dynamic synergism bonding

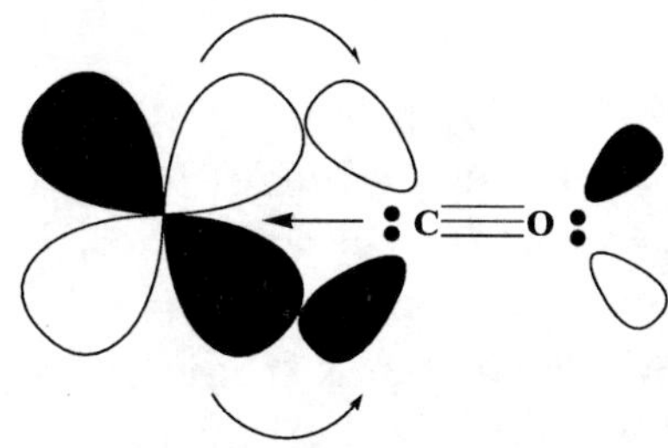

9.   M—C≡O: ⟷ M=C=O:

10. Iodide is a larger π-donor ligand

11. σ-donor ligand and π-acceptor

12. The water molecule has an oxygen atom with more than one pair of electrons on it, so it can function as a π-donor ligand, which decreases the field strength. Also, ammonia is a stronger Lewis base and a better sigma donor than water.

13. The o-phenanthroline (phen) ligand has the possibility of forming π-acceptor bonding interactions with the metal, whereas ammonia can not.

14. trimethyl phosphite

15. The compound must exhibit the structure shown in order for the alkene to use its pi* orbitals to effectively backbond to the filled platinum d-orbitals.

Sigma Bond Component   Pi Acceptor Bond Component

Chapter 7 Answers to Problems and Exercises

1. [V(CO)$_6$], [Cr(CO)$_6$], [Fe(CO)$_5$], [Ni(CO)$_4$].
All of these compounds satisfy the eighteen electron rule,
the noble gas formalism, except for the vanadium
compound. Vanadium would have a total of five
electrons, and with the twelve electrons from the carbon
monoxide ligands the total would be therefore 17e-'s, and
thus vanadium carbonyl would be a paramagnetic
compound.

2. With an uneven number of valence electrons the
metals form a metal-metal bond in order to satisfy the
octet rule and become diamagnetic, and not remain
paramagnetic with one unpaired electron.

3.

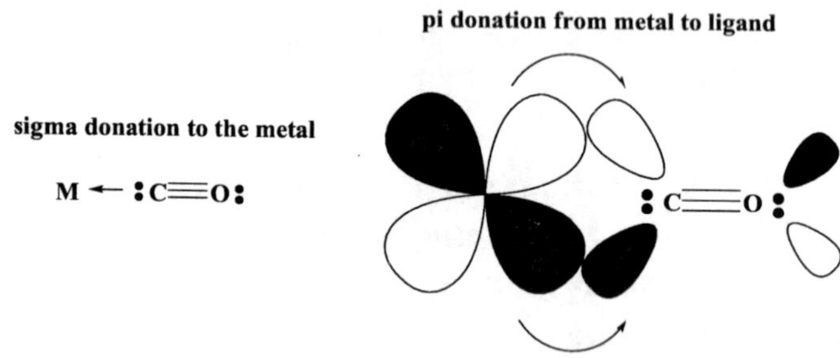

4.

Terminal,   bridging,   multiple metal bridging,   semi-bridging

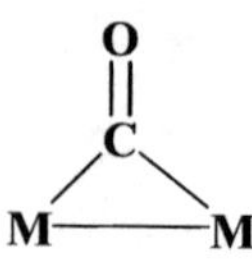

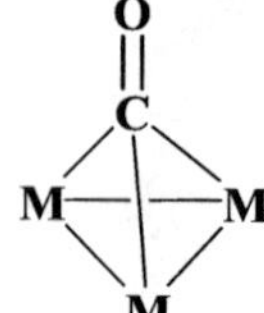

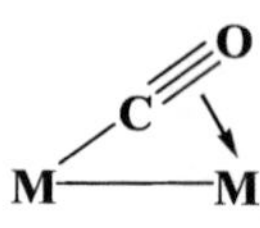

5. The more negative charge on the metal center (or the more electron density on the metal center) results in increased backbonding from the metal to the ligand. The more backbonding to the ligand from the metal center means that the antibonding molecular orbitals on the carbon monoxide ligand become more populated, which results in decreased bonding between the carbon and oxygen atoms in the CO molecule. When the bond order is decreased from three to somewhere between three and two, then the infrared stretching frequency is reduced as well. Thus, the $V(CO)_6^-$ molecule has the most negative charge on the metal, the most backbonding, and therefore the lowest stretching frequency. The $Mn(CO)_6^+$ molecule has the most positive charge on the metal, the least backbonding, and therefore the highest stretching frequency.

6. (a) $Fe_2(CO)_9$,

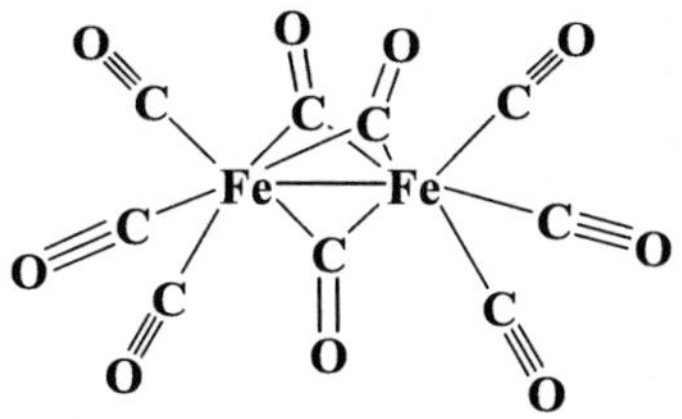

(b) $Ru_3(CO)_{12}$

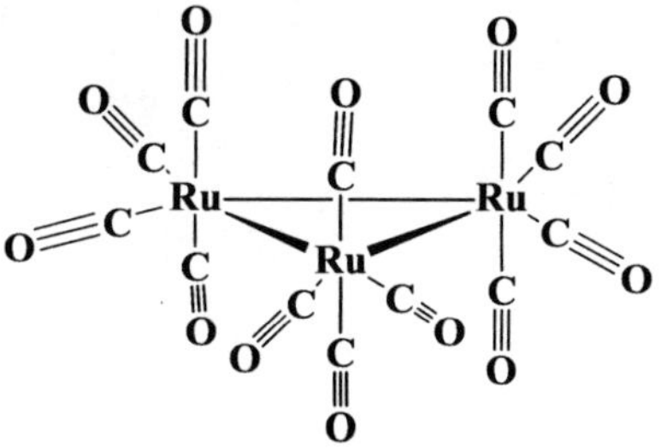

(c) $Rh_4(CO)_{12}$

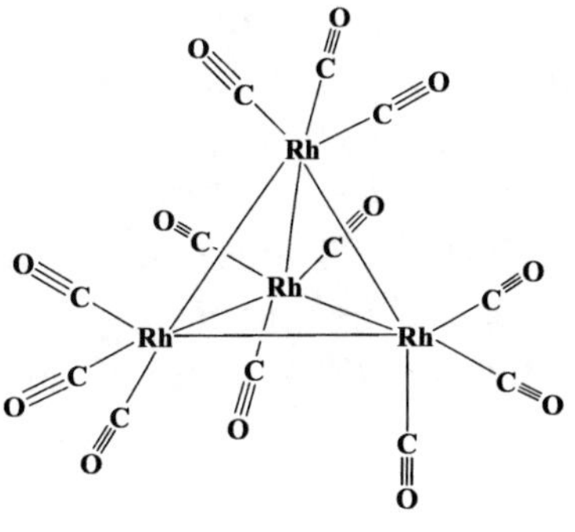

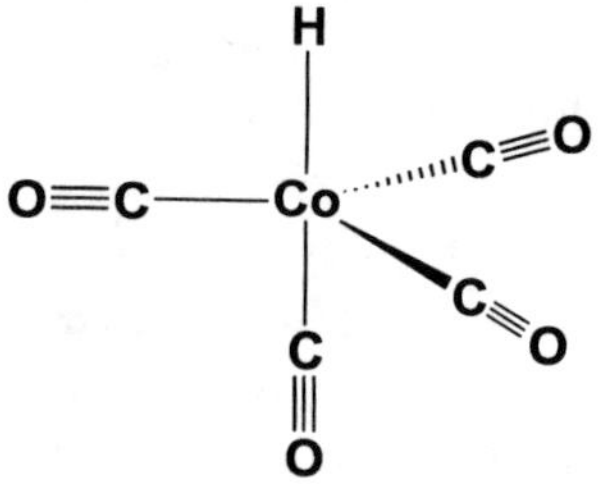

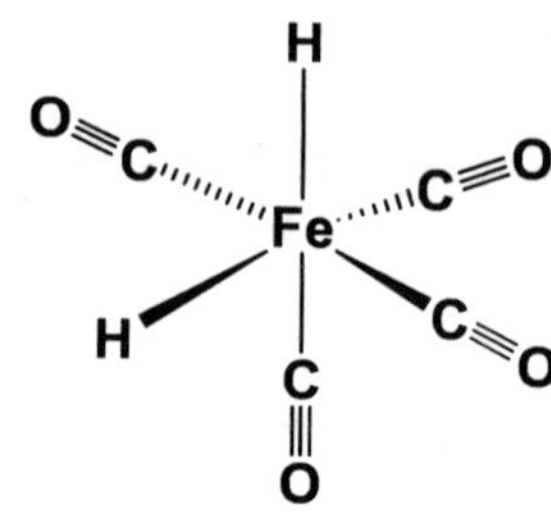

4 Terminal CO   8e-

Terminal H ⁻   2e-

Co⁺   8e-

____

18e-

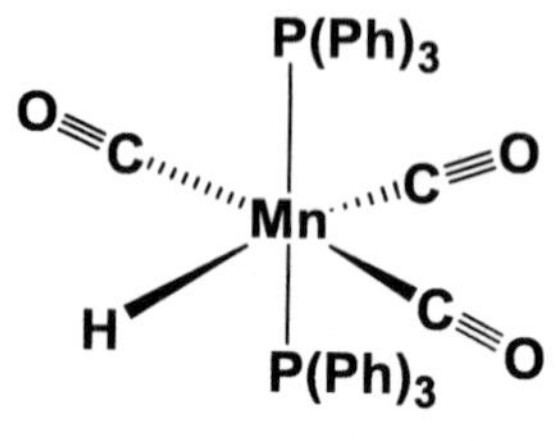

4 Terminal CO   8e-

2 Terminal H ⁻   4e-

Fe²⁺   6e-

____

18e-

9.

3 Terminal CO   6e-
2 Terminal P(Ph)₃   4e-
1 Terminal H⁻   2e-
Mn⁺   6e-

____

18e-

10. The Mn group has an odd number of valence electrons, seven, and thus it needs to form a metal-metal bond (or become oxidized to-for example Mn+) in order to achieve eighteen electrons. The compound $Mn(CO)_6$ for example would have a total of 19-electrons. Also, $Mn(CO)_5$ would have a 17-electrons center. By forming a metal-metal bond, and coordinating five carbonyls each, the dimer would achieve 18-electrons around each metal.

11.

(a)

2 Triphenyl phosphite	4e-
4 Terminal CO$^-$	8e-
Mo	6e-
	18e-

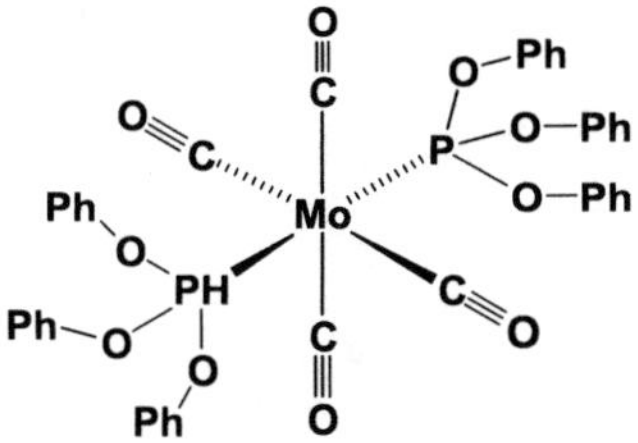

(b)

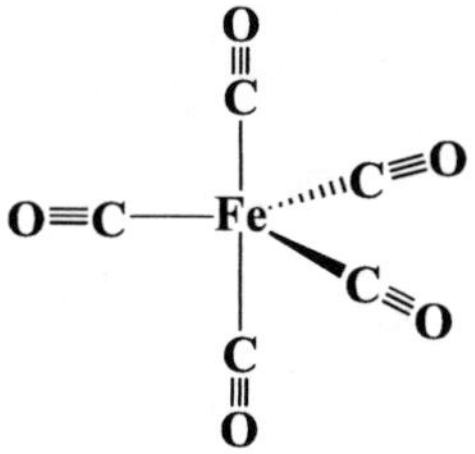

Fe	8e-
5 terminal CO	10e-
	18e-

(c)

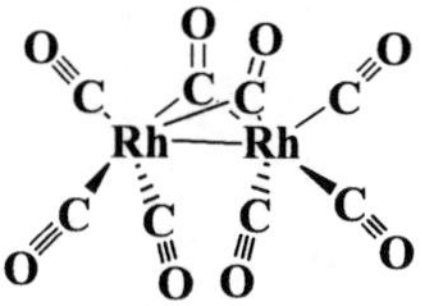

Rh	9e-
M-M bond	1e-
3 terminal CO	6e-
2 bridging CO	2e-
	18e-

(d)

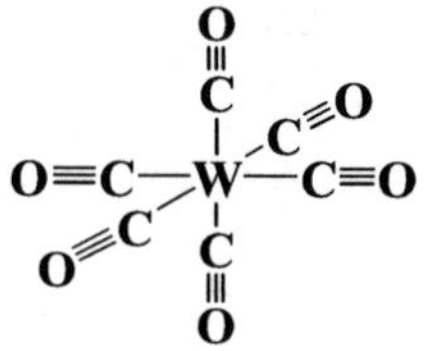

W	6e-
6 terminal CO	12e-
	18e-

(e)

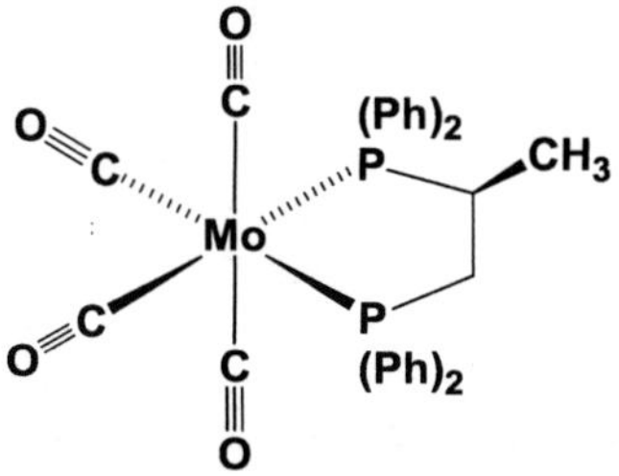

PROPHOS	4e-
4 Terminal CO$^-$	8e-
Mo	6e-
	18e-

(f)

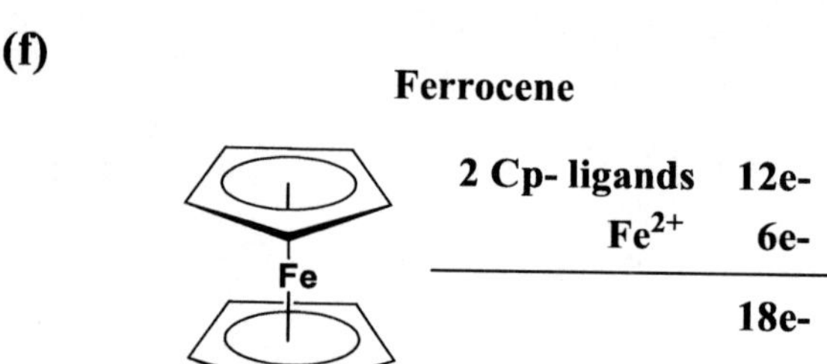

Ferrocene

2 Cp- ligands	**12e-**
Fe^{2+}	**6e-**
	18e-

(g)

Wilkinson's Catalyst

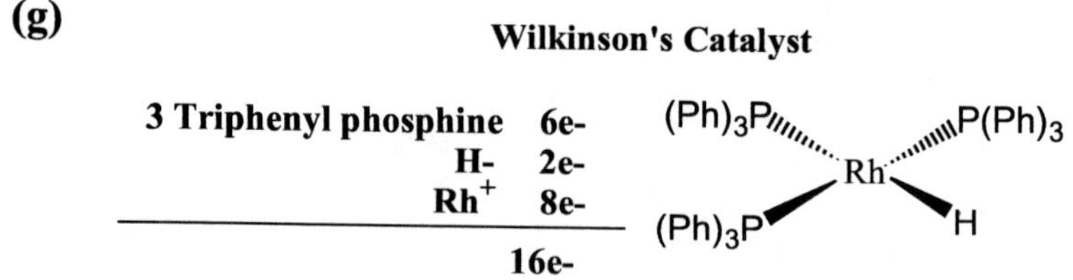

3 Triphenyl phosphine	**6e-**
H-	**2e-**
Rh$^+$	**8e-**
	16e-

(h)

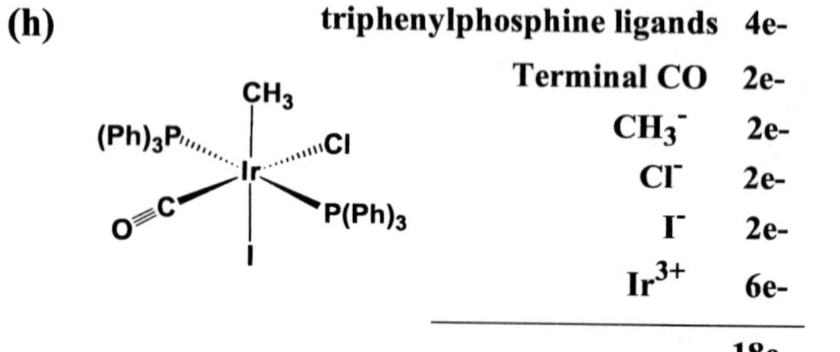

triphenylphosphine ligands	**4e-**
Terminal CO	**2e-**
CH$_3^-$	**2e-**
Cl$^-$	**2e-**
I$^-$	**2e-**
Ir^{3+}	**6e-**
	18e-

(i)

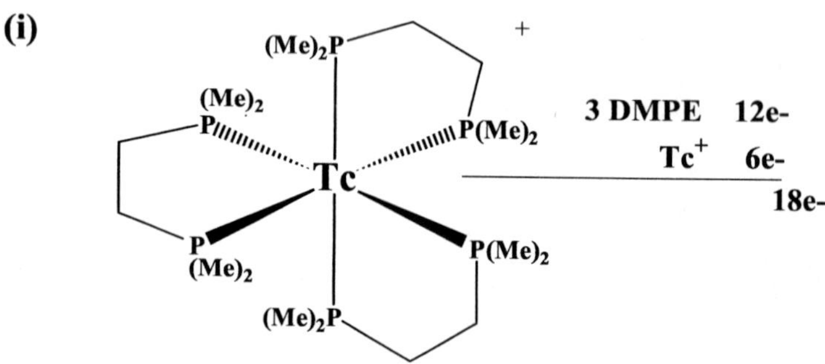

3 DMPE	**12e-**
Tc$^+$	**6e-**
	18e-

*two isomers are possible (lambda and delta—see page 108)

(j)

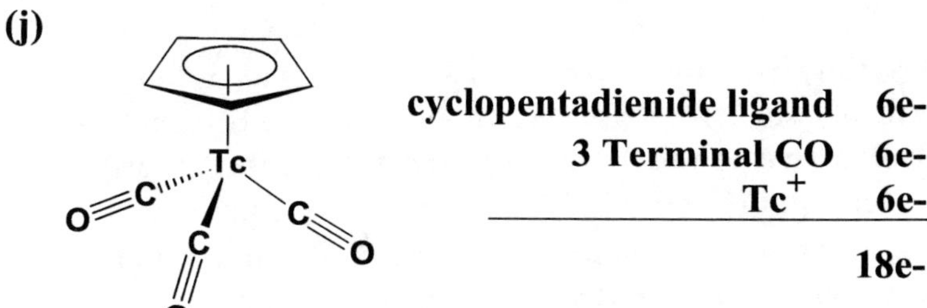

cyclopentadienide ligand	6e-
3 Terminal CO	6e-
Tc^+	6e-
	18e-

(k)

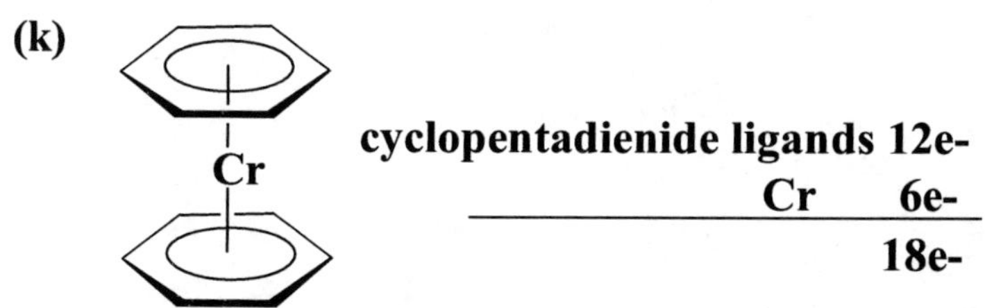

cyclopentadienide ligands	12e-
Cr	6e-
	18e-

(l) **Each metal has to be counted separately**

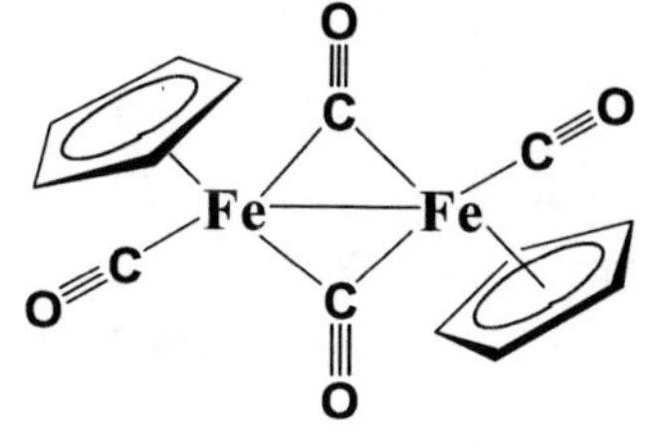

cyclopentadienide ligands	6e-
one terminal CO ligand	2e-
two bridging CO ligands	2e-
metal-metal bond	1e-
Fe+	7e-
	18e-

12.

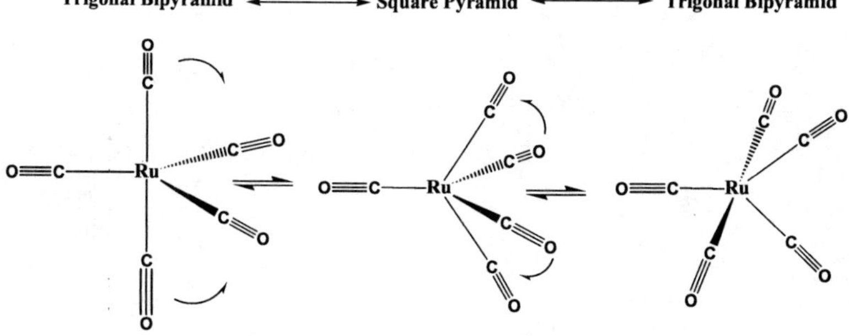

13.

(a) Synergism comes from the Greek word "*synergos*" meaning working together. It refers to the interaction between two or more things when the combined effect is greater than if you added the things together.

(b) Increased backbonding reduces CO bond order, decreases the infrared stretching frequency.

14.

(a)

$$Co_2(CO)_8$$

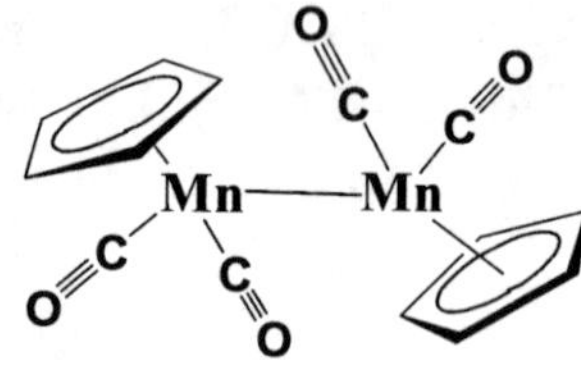

Co	9e-
M-M bond	1e-
3 terminal CO	6e-
2 bridging CO	2e-
	18e-

(b)

Each metal has to be counted separately

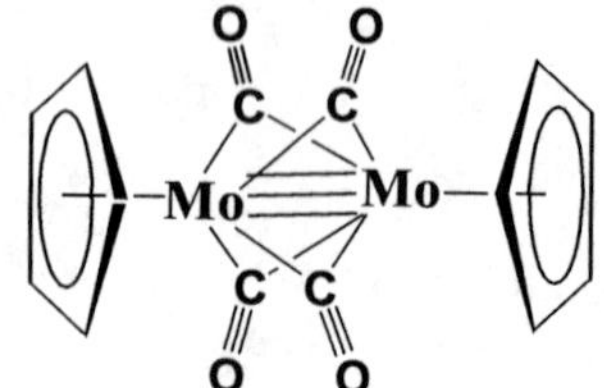

cyclopentadienide ligand	6e-
two terminal CO ligands	4e-
two metal-metal bonds	2e-
Mn+	6e-
	18e-

(c)

Each metal has to be counted separately

cyclopentadienide ligand	6e-
four bridging CO ligands	4e-
three metal-metal bonds	3e-
Mo+	5e-
	18e-

15. **A quadruple bond is present.**

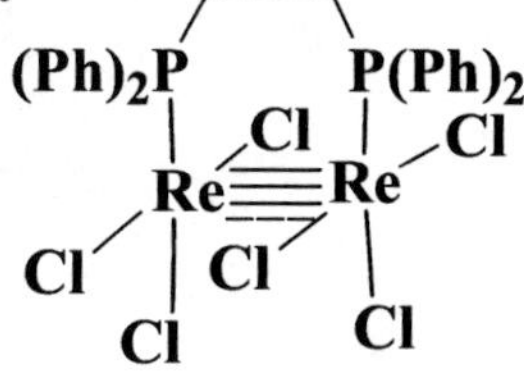

16.

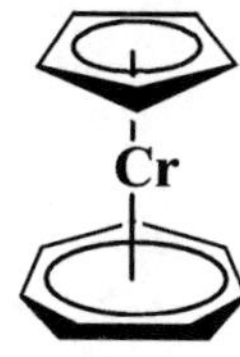

tropylium$^+$ ligand	6e-
cyclopentadienide ligand	6e-
Cr^0	6e-
	18e-

benzene ligand	6e-
cyclopentadienide ligand	6e-
Mn$^+$	6e-
	18e-

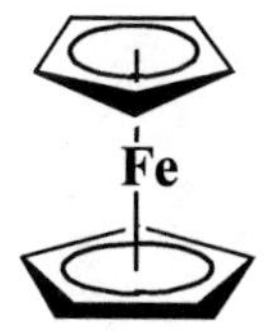

cyclopentadienide ligands	12e-
Fe^{2+}	6e-
	18e-

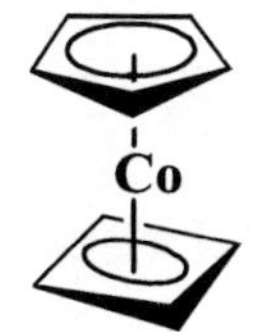

cyclobutadiene ligand	4e-
cyclopentadienide ligand	6e-
Co$^+$	8e-
	18e-

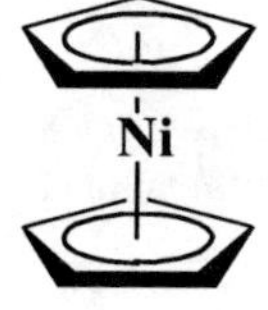

cyclopentadienide ligands	12e-
Ni^{2+}	8e-
	20e-

17.

2 Terminal P(Ph)₃   4e-
1 Terminal NO+   2e-
1 Terminal Cl⁻   2e-
Ru   8e-
16e-

3 Terminal CO   6e-
1 Terminal NO+   2e-
Co⁻   10e-
18e-

2 Terminal CO   4e-
2 Terminal NO+   4e-
Fe²⁻   10e-
18e-

1
3 Terminal CO   6e-
2 Bridging CO   2e-
2 Metal-Metal bonds 2e-
Fe   8e-
18e-

2
3 Terminal CO   6e-
2 Bridging CO   2e-
2 Metal-Metal bonds 2e-
Fe   8e-
18e-

3
4 Terminal CO   8e-
2 Metal-Metal bonds 2e-
Fe   8e-
18e-

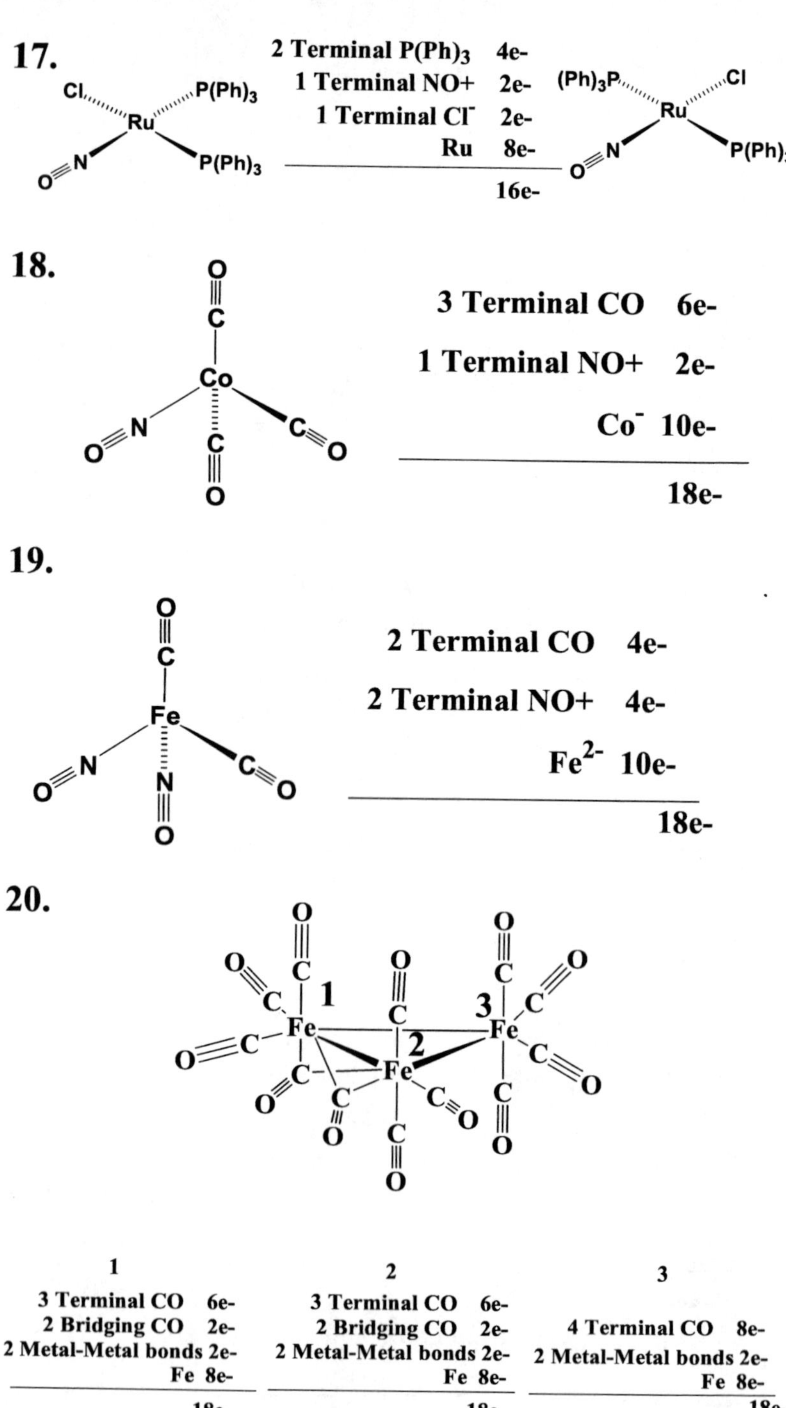

18.

19.

20.

Notes